Zelalem Yilma

Halloumi Cheese

Sara Endale
Mitiku Eshetu
Zelalem Yilma

Halloumi Cheese

Evaluation of Bovine, Caprine and Ovine Rennet Types on Yield and Quality of Halloumi cheese

LAP LAMBERT Academic Publishing

Impressum / Imprint

Bibliografische Information der Deutschen Nationalbibliothek: Die Deutsche Nationalbibliothek verzeichnet diese Publikation in der Deutschen Nationalbibliografie; detaillierte bibliografische Daten sind im Internet über http://dnb.d-nb.de abrufbar.
Alle in diesem Buch genannten Marken und Produktnamen unterliegen warenzeichen-, marken- oder patentrechtlichem Schutz bzw. sind Warenzeichen oder eingetragene Warenzeichen der jeweiligen Inhaber. Die Wiedergabe von Marken, Produktnamen, Gebrauchsnamen, Handelsnamen, Warenbezeichnungen u.s.w. in diesem Werk berechtigt auch ohne besondere Kennzeichnung nicht zu der Annahme, dass solche Namen im Sinne der Warenzeichen- und Markenschutzgesetzgebung als frei zu betrachten wären und daher von jedermann benutzt werden dürften.

Bibliographic information published by the Deutsche Nationalbibliothek: The Deutsche Nationalbibliothek lists this publication in the Deutsche Nationalbibliografie; detailed bibliographic data are available in the Internet at http://dnb.d-nb.de.
Any brand names and product names mentioned in this book are subject to trademark, brand or patent protection and are trademarks or registered trademarks of their respective holders. The use of brand names, product names, common names, trade names, product descriptions etc. even without a particular marking in this work is in no way to be construed to mean that such names may be regarded as unrestricted in respect of trademark and brand protection legislation and could thus be used by anyone.

Coverbild / Cover image: www.ingimage.com

Verlag / Publisher:
LAP LAMBERT Academic Publishing
ist ein Imprint der / is a trademark of
OmniScriptum GmbH & Co. KG
Heinrich-Böcking-Str. 6-8, 66121 Saarbrücken, Deutschland / Germany
Email: info@lap-publishing.com

Herstellung: siehe letzte Seite /
Printed at: see last page
ISBN: 978-3-659-62675-3

EVALUATION OF BOVINE, CAPRINE AND   OVINE RENNET TYPES ON YIELD AND QUALITY OF HALLOUMI CHEESE

SARA ENDALE

ACKNOWLEDGEMENT

First of all I would like to express my earnest gratitude to my major advisor Dr. Mitiku Eshetu for his invaluable professional advice and unreserved fatherly approach throughout the course and this thesis work. My sincere appreciation extends to my co-advisor Dr. Zelalem Yilma for his invaluable comment, technical advice and support provided at the time of the research as well as during thesis write-up process. His collaboration is really acknowledged.

I would like to thank my employer, Dilla University for sponsoring my in-country scholarship for the M.Sc. study. My special thanks also go to Holeta Agricultural Research Center for supporting my project. My heartfelt thanks and deepest appreciation go to Rahel Nebiyu and Binyam Kassa, who gave me valuable comments and constructive ideas that shaped this study from research topic selection to thesis write up. A note of thanks is also extended to Ato Wondimeneh Esatu, Kassahun mellese, Dr. Keffena Effa and Yalembirhan Molla for their support during place arrangement.

I deeply thank Zewdie Wondatir and Dr. Mengistu Alemayehu for their support especially in statistical analysis. My special thanks also go to Alemitu Beyene,

Kassahun Yemane, Yosef Kehalew, Banchayew for their support during Laboratory work. I am also thankful to Mekedes Mengeiste, Mebrat Tola, Mahilet Mekonnen, Betelihem Gemechu, Yohannes Kebede, Selamawit Girmay, Eyerusalem Tesfaye and worknesh seid for their friendship collaboration and help especially during the study time. You all made my stay at Haramaya more enjoyable.

I cherish the love and moral support I received from my Mother Tsiga G/Yes and my Brother Behailu Endale (mamush) throughout my life and this thesis would not be a reality if it was not for the love and moral support of my family. I want to forward my special gratitude to my boyfriend Eyob Goshu for his encouragement to withstand the hardships of the graduate study with all those pains I faced. I thank you All!

Last, but not least, I thank the Almighty God for His unconditional love, protection and bringing all the aforementioned people into my life.

ACRONYMS AND ABBREVIATIONS

ANOVA	Analysis of Variance
AD	Anna Domine
β-casein	Beta casein
BCFA	Branch chain fatty acid
CCP	Colloidal calcium phosphate
Cfu	Colony forming unit
CEC	Cation exchange capacity
CRD	Completely randomized design
CSA	Central Statics Agency
CV	Coefficient of variation
Ca^{+2}	Calcium ion
$Cacl_2$	Calcium chloride
CC	Coliform count
EIAR	Ethiopian Institute of Agricultural Research
ETB	Ethiopian Birr
FAO	Food and Agriculture Organization of the United Nations
FAOSTAT	Food Aid Organization Stastical Division

FPC	Fermentation produced chymosin
GMO	Genetically modified organism
HARC	Holetta Agricultural Research Center
IMCU	International Milk Clotting Unit
k-casein	Kappa casein
Kg	Killo gram
LAB	Lactic acid bacteria
LSD	Least significant difference
Leu	Leucine
MCA	Milk clotting activity
Met	Methonine
MOA	Ministry of Agriculture
m.a.s.l	Meter above sea level
Ml	Mili liter
NaOH	Sodium hydro oxide
Nacl$_2$	Sodium Chloride
OM	Organic matter
PCA	Plate count agar
pH	Power of Hydrogen

Phe	Phenylalanine
Ppm	Part per million
RCT	Rennet clotting time
RU	Rennet Unit
SAS	Statistical analysis software
S.E	Standard Error
SU	Soxhlet unit
TPC	Total plate count
Tyr	Tyrosine
TS	Total solid
USD	United states dollar
Val	Valine
VRBA	Violet red bile agar
w/v	Weight by volume
YMC	Yeast and mould count

DEDICATION

I dedicate this thesis document to best mother of the world, Tsiga G/yes and My Brother Behailu Endale for their affection, love and dedicated partnership in the success of my life.

TABLE OF CONTENTS

1. INTRODUCTION

Cheese is the curd or hard substance formed by the coagulation of milk of certain mammals by rennet or similar enzymes in the presence of lactic acid produced by added or adventitious microorganisms from which part of the moisture has been removed by cutting, warming and/or pressing, which has been shaped in a mould and then ripened by holding for sometime at suitable temperatures and humidity (Castillo, 2001).

Cheese is made in almost every country in the world with the existence of more than 2,000 varieties (O'Connor, 1993). It provides an ideal medium for preservation of valuable nutrients in milk and is an excellent source of protein, fat, minerals, vitamins, and essential amino acids.

Halloumi cheese is a semi hard to hard unripened elastic cheese with no obvious skin/rind and a texture that is close with no holes and it can easily be sliced. Its color varies from white (when produced from sheep's or goat's milk) to yellowish (when produced from cow's milk) (Papademas and Robinson, 1998). Traditionally, Halloumi cheese was consumed in Cyprus, but today it has become a very popular cheese in many countries around the world (Papademas, 2006). The properties of Halloumi cheese and its manufacture have been reviewed and studied by different researchers (Papademas and Robinson, 1998, 2000; Milci *et al.*, 2005; Raphaelides *et al.*, 2006; Papademas, 2006; Kaminarides *et al.*, 2007).

Traditionally, Halloumi cheese used to be produced only from raw ovine or caprine milks, alone or as blends. Due to the growing market demand, new regulations were created to permit the use of bovine milk if the origin of the cheese milk is declared on the label of the product (Moatsou *et al*, 2004). Bovine milk offers remarkable advantages over the traditional ovine and caprine milks by being less expensive and more available. Although the chemical composition of milk from the three sources differs significantly, the shift to the use of bovine milk in the production of Halloumi has been an interesting economical success (Papademas and Robinson 1998; Papademas, 2006).

Demand for cheese is showing an increasing trend due mainly to the change in feeding habits of a substantial proportion of consumers. One shift worth mentioning is increasing demand for fast foods, in which cheeses such as Halloumi are constituents, driven by life style fast foods are more practical for most employees and students.

Success in making a cheese acceptable in both flavor and texture depends partly on curd properties which determine the retention of fat and moisture and, thus, cheese yield and composition (Green and Grandison, 1993). An important factor affecting cheese yield and quality is the rennet used in production.

Milk-clotting enzyme is the first active agent in cheese-making and essential ingredients for the production of various cheese varieties. Clotting of milk is usually an enzyme-driven step during which rennet enzymes bring about cleavage of Met105– Phe106 of k-casein. Today, nearly all types of cheese are manufactured by clotting the milk with a milk clotting enzyme and then processing the curd in various ways. For this purpose, proteolytic enzymes originating from plants, animal tissues and microorganisms are widely used. The use of rennet in cheese making dates back to approximately 6000 BC (Moschopoulu, 2011).

For centuries, calf rennet has been used as a milk coagulant in the production of all varieties of cheese. According to the report of FAO (2010), since 1961, the world cheese production increased by a factor of approximately 3.5 with a decrease in the use of rennet due to limited availability of calf stomachs. This led to the need to search for substitutes (Jacob *et al.*, 2011). Demand for coagulating enzymes started exceeding the supply already about 40 years ago (de Koning, 1978) and today, only 20-30% of the demand for coagulants can be covered by calf rennet (Jacob *et al.*, 2011).

Although many proteinases can induce the coagulation of milk, most are too proteolytic or have the incorrect specificity and hence can lead to reduced cheese yield or defective cheeses. The clotting activity of such proteinases can also affect curd properties, such as firmness or softness, during processing (Fox and Stepaniak, 1993). Thus, for both economic and processing reasons, it is essential to know the milk clotting activity of certain rennet type, which allows the cheese maker to select the adequate enzyme for each type of cheese and adjust enzyme proportion to control milk coagulation and thus optimize cheese yield and quality.

Considering the nutritional, economic and value addition importance of cheeses as well as the unaffordable price of commercially available rennet to most resource poor producers; it is essential to search for locally available and affordable rennet. In Ethiopia, due to the high cost, only a few dairies manufacture cheese using commercial rennet to supply mainly to large pizza house, Hotels and supermarkets. Still compared with some related commercial cheese types, Halloumi cheese is cheaper, and also has higher nutrient content and better keeping quality (O'connor, 1993; Papademas, 2006). As there is no need of starter culture to manufacture Halloumi cheese, its production cost is lower.

Each year in Ethiopia, a significant number of livestock are slaughter for domestic meat consumption as well as export. According to the annual report of the Central Statistics Agency of Ethiopia, for instance, 340000 cattle, 3.08 million sheep, and 1.77 million goats were slaughtered in Ethiopian year just ended (2012/2013) (CSA, 2013). This means that, a substantial volume of liquid rennet can be prepared from adult livestock. However, there is no or very limited information available on production, economics and efficiency of rennet making from locally available sources. This prevailing situation makes the basis of this M.Sc. research work, which was conducted with the following objectives:

➢ To assess options of rennet making from locally available sources (abomasums of mature bovine, caprine and ovine)

➢ To compare the strength and shelf life of bovine, caprine and ovine rennet types with commercial ones

➢ To evaluate the suitability of locally available liquid rennet extracted from mature bovine, ovine and caprine for Halloumi cheese making

7

2. LITURATURE REVIEW

2.1 Milk Production in Ethiopia

Ethiopia has one of the largest livestock inventories in Africa with a national herd estimated at about 54 million cattle, 25.5 million sheep, 24 million goats, 9 million pack animals (donkeys, horses and mules) and 916 thousand camels in 2012/13 (CSA, 2013). In the same reporting year, total female cattle constituted for 55.48% of the total national cattle population. Of the total female cattle population, dairy cows totaled at 6.75 million (12.5%) with the proportion of hybrid and pure exotic female cattle population being only 0.69% (CSA, 2013).

In 2012/13, the total cow milk production at country level was estimated at about 3.8 billion liters with average daily milk yield per cow of about 1.32 liters (CSA 2013). In the same reporting year, camels produced about 165 million liters of milk with daily average milk yield per camel being 3.56 liters. Considering the annual milk produced by cows and camels, cows accounted for about 95.8 percent with the remainder about 4.2% being produced by camels. The majority of milking cows are indigenous breeds with low production performance. On average, zebu cattle breeds in the high and mid altitude areas of Ethiopia do not yield more than one kg of milk per cow per day (Tesfaye, 1992). However, crossbred dairy cows could yield 4-5 times the milk obtained from the indigenous cattle (Tesfaye, 1995).

In Ethiopia there are three major types of milk production systems (Tsehay, 2002); the urban milk production system (the city of Addis Ababa, and regional towns), the peri-urban milk production system (proximity to Addis Ababa and regional towns) and the rural milk production system (farmers in the villages). According to Azage and Alemu (1998), almost all of the fluid milk supplied to major urban centers in the country comes from urban and peri-urban dairy producers. The milk produced in most of the smallholder farms is either sold and/or consumed as fresh, fermented milk and products such as butter and cheese (O'Connor, 1994). Milk and milk products have important role in feeding the rural and urban population of Ethiopia owing to their high nutritional value.

According to recent studies the total milk production is increased significantly (300%) from what it has been the year 2000 (Haile, 2009). Given the high potential for dairy development and the ongoing policy reforms and technological interventions, success similar to that realized in the neighboring Kenya under a very similar production environment is expected in Ethiopia (Ahemed *et al.*, 2004).

2.2 Milk and Milk Products Consumption in Ethiopia

Milk and milk products form part of the diet for many Ethiopians. They consume dairy products either as fresh milk or in fermented or soured form. According to Getachew and Gashaw (2001), 68% of the total milk produced is used for human consumption in the form of fresh milk, local butter, Ergo (Ethiopian naturally fermented milk) and Ayib (Ethiopian acid-heat treated soft type cottage cheese) while the rest is given to calves and wasted in the process.

Ethiopia consume less dairy products than other African countries and far less than the world consumption. The present national average capita consumption of milk is 19kg/year as compared to 27kg for other African countries and 100kg to the world per capita consumption (FAO, 2003). Consumption of milk products in Ethiopia is associated with income level, availability, and cultural restriction. Dairy products are not consumed during fasting seasons, and also on Wednesday and Friday among the followers of Ethiopian Orthodox Church. Consumption is also restricted during dry seasons, and in areas where dairy products are in short supply because of the critical shortage of feed for the dairy animals (Kassahun, 2008).

The consumption of milk and milk products vary geographically between the highlands and low lands and level of urbanization. In the lowlands, all segments of the population consume dairy products while in the highlands major consumers include primarily children and some vulnerable groups of women (Ahmed *et al.*, 2003).

These days there is rapid population growth and urbanization in the country. Moreover, the income of the urban dwellers is growing. These factors have played an important role to increase the demand for dairy products especially in urban areas. The demand for dairy products depends on consumer preference, consumer's income, population size, price of the product, price of substitutes and other factors. Increasing population growth, rising real income and decreasing consumer prices are the major factors that are expected to increase the demand for dairy products (Ahmed *et al.*, 2004).

2.3 Milk Processing Practices in Ethiopia

Generally, people in the tropics have mastered the art of conserving or processing the limited quantity of the milk they produce due to the generally prevailing unorganized processing, transport and marketing facilities (Taye, 1998).

A large amount of milk produced in tropical countries is converted into indigenous products such as butter and cottage type cheese or some kinds of fermented or concentrated product that have longer shelf life (Chamberlain, 1990). In rural areas, milk may be processed fresh or sour (O'Connor, 1994). The choice depends on available equipment, product demand and on the quantities of milk available for processing.

In many parts of Ethiopia, the climate is hot for most of the year and under such conditions the raw milk spoils easily during storage at ambient temperatures. As reported by O'Mahoney and Peters (1987), cooling systems are not common place in many parts of the rural Ethiopia. In addition, there are post harvest losses associated with poor handling, contamination, low level of technology applied in the conservation of milk to extend its shelf life and limited access to milk market (Felleke, 2003).

Increasing shelf life, and in some cases surplus production, are the reasons for milk processing (Zelalem and Inger, 2000). In the country, milk is processed into different dairy products either on large-scale commercial dairies or smallholder farms. Traditional processing technologies are generally considered to be time consuming

11

and inefficient in terms of milk fat recovery, product quality, a comparatively short shelf life and provide little return for the milk producer (Felleke, 2003).

An understanding of the farmers' products and the conditions under which they operate are among the prerequisites for improvement in the efficiency of the locally available technologies and/or adoption of new technologies by the small farmers (Fekadu, 1994).

2.4 Cheese Production and Utilization

A third of the world's milk production is used to produce cheese that is of great diversity of flavors, textures and forms (Fox, 2003). Despite the large number of verities (2000) cheese may be classified into different groups, such as ripened and un ripened cheese, cheese with low or high fat content and cheese with soft or hard consistency (O'Connor, 1993).

In countries with a well developed dairy industry, cheese provides an ideal vehicle in preserving the valuable nutrients of milk. Cheese is an excellent source of protein, fat and minerals such as calcium, iron and phosphorous, vitamins and essential amino acids, making it an important food in the diet of both young and adult human beings (Raheem, 2006).

Recipes and processes are passed from generation to generation by observation and practical experience. Even though traditional cheese varieties are produced at a small-scale level using indigenous technologies, they serve as a valuable source of nutrients and income for many small producers. However, insufficient amount of cheese is manufactured to meet the demand (Walstra, 1993).

The demand for dairy products in Sub-Saharan Africa continues to increase, with the overall growth rate in the consumption of milk and milk products estimated at about 2.1% per annum (Raheem, 2006). The growth in demand results from rapidly rising populations, urbanization and some increase in per capita income. This increasing demand for dairy products offers a great opportunity and potential for the low output dairy producer and provides an incentive for the development of dairy processing industry (O'Connor, 1993).

2.5. Halloumi Cheese

The majority of white brined Halloumi cheese is rennet coagulated cheese (Toufeili *et al.*, 2006). The gross chemical composition of Halloumi cheese was not only affected by the type of milk used but also by the method of manufacture (Papademas, 2006).

2.5.1. Historical background

Halloumi cheese is the traditional cheese of Cyprus. Legends suggest that the cheese was introduced to the island by Arab mercenaries from Syria and Palestine during the Frankish rule (AD 1192-1489). The oldest record that refers to the production of Halloumi in Cyprus dates back to 1554 when the Italian writer Florio Bustron mentioned a cheese named Halloumi produced from a mixture of sheep's and goat's milk (Papademas, 2006). Since then, the production of Halloumi cheese has flourished, notably in Cyprus. According to Gibbs and Morphitou (2004), Halloumi cheese is not only consumed in Cyprus but it is also exported to many countries.

2.5.2. Types of halloumi cheeses

There are two types of Halloumi cheese grouped based on maturity and texture: fresh and mature.

Fresh Halloumi Cheese: This is the type consumed directly after manufacture or in other words, twenty four hours after production. Fresh Halloumi cheese has a distinct aroma and a mild milky and creamy flavor. The texture is close with no holes and it is easily sliced. Fresh Halloumi cheese is usually described as moderately salty (Papademas, 2006).

Mature Halloumi Cheese: Mature Halloumi cheese is the product that is matured for at least 40 days after production before being introduced to the market. Alterations in the texture and flavor of the cheese take place during maturation. The most evident change is the significant increase in hardness. Accordingly, mature Halloumi cheese has a tougher body and a closer texture than fresh cheese (Papademas, 2006).

2.5.3. Chemical composition

The gross chemical composition of the fresh and mature Halloumi cheeses made from different milks and methods of manufacture is different (Kaminarides et al., 2000; Papademas, 2000). Fresh Halloumi cheese is reported to contain around 45.65% water, 22.81% protein, 25.43% fat and 6.11% ash (Biilent et al., 2011). Papadidmes et al., (2000) also reported that fresh Halloumi cheese made from full fat comprises about 45% moisture, 28% fat, 21.1% protein. Mutamed et al. (2007), on the other hand, reported 50.48% moisture, 20.13% protein, 21.30% fat and 3.86% ash for Fresh Halloumi cheese. According to Biilent et al. 2001, the average pH of Fresh Halloumi cheese ranged from 5.37–6.45.

2.5.4. Microbiological quality

Although the high temperatures employed to pasteurize raw milk and cook the Halloumi curd destroy most microorganisms, Halloumi cheese is by no means free from microorganisms. Several studies have been conducted to compare and identify microorganisms isolated from fresh and mature Halloumi cheese produced from different type and age types of milks (Papademas and Robinson, 1998: Milci et al., 2005: Papademas and Robinson, 2000: Bintsis et al., 2000).

Papademas and Robinson (2000) reported that the microbiological quality of commercial Halloumi cheeses (fresh and matured) was different, mainly due to type of the product. The major difference between the fresh samples of Halloumi cheese was the apparent absence of lactic acid bacteria (LAB) in the industrial samples. Low levels of LAB were observed in both types of cheese due to the high temperatures employed for raw milk pasteurization and curd cooking.

Yeast isolated from fresh Halloumi cheese have probably originated from the dairy's environment and/or from the whey brines used to immense the blocks for short period before they are packaged and sold as fresh product (Bintsis and Papademas, 2002). The yeast isolated from the commercial samples of fresh and mature Halloumi cheese did not have any adverse effect on the flavor or produce visible defects.

In addition, post pasteurization contaminants could survive in brine solution. The use of high salt solution to eliminate pathogens has been considered in Halloumi cheese. However, it should be noted that around such salt concentration flavor is negatively affected and regulation in limit the salt concentration in Halloumi cheese to 3% as a maximum (Papademas and Robinson, 1998).

2.5.5. Sensory profiling and rheological properties of halloumi cheese

Organoleptic quality is a term pertaining to the sensory properties of a particular food or chemical (Lawless and Heymann, 1999). Any food product should have desirable appearance to the senses of the consumers in order to get in the transaction. Such properties of Halloumi cheese are briefly discussed below.

Flavor and Taste: Halloumi cheese, as with all other cheese types, has a unique flavor and taste. A study by Papademas & Robinson (2000) showed that the organoleptic characteristics of Halloumi cheese change dramatically with maturation, and it becomes hard in texture, and salty and acidic in taste, while flavor notes such as 'creamy' and 'milky' decrease. The study of the flavor compounds of Halloumi cheese has revealed great differences both with age and between samples. The major flavor compounds representing this volatile fraction were probably formed by the action of heat (lactones and methyl ketones in fresh cheese), lipolysis and other chemical reactions, including reductions and oxidations. Short-chain organic acids (i.e. ethanoic, propanoic) may contribute to the flavor of cheeses with low levels of proteolysis and lipolysis (de Llano *et al.*, 1996), while the branch-chain fatty acids (BCFA) are considered important flavor constituents due to their 'cheesy' flavor notes (Brennand *et al.*, 1989).

Texture: Milk fat serves multiple functions in a variety of cheeses. Its impact on various physical properties such as firmness, adhesiveness and mouth feel are evident when fat is removed from the cheese (Norman *et al.*, 1990). In a study conducted by Papademas *et al.* (2000), milk fat was found to affect the meltability and stretchability of Halloumi produced from bovine milk. Full fat cheese (28 percent fat) showed significantly more melting and stretching properties than low fat cheese (19.3 percent fat) and reduced fat cheese (10.8 percent fat).

The texture of fresh and mature Halloumi cheese has been studied extensively, and the results showed that the most pronounced effect of maturation was the significant increase in hardness, detected in all samples. The fracturability, as perhaps expected, also increased during maturation. Differences were noted between samples of fresh Halloumi; the industrial Halloumi cheese (bovine milk as the major constituent) was the hardest, the most chewy and least fracturable, and the traditional product (mixture of sheep's and goat's milk) was the least hard, least chewy and most easily fractured. The fresh samples did not differ significantly as far as springiness and cohesiveness were concerned (Papademas, 2000).

The increase in hardness during maturation of the cheese is probably a direct impact of the penetration of salt into the product from the whey brine, and the concomitant loss of water into the whey brine. It has also been proposed by Walstra *et al.* (1999) that, the quantities of water lost and amount of salt uptake are strongly positively correlated with the duration of brining and the salt content of the brine. In addition, Kaminarides & Anifantakis (1985) reported a loss in weight during the maturation of Halloumi cheese. The chemical composition data of the commercial Halloumi samples reveal a relationship between the texture of the cheeses and maturation, as the moisture content decreased, the salt content increased.

2.5.6. Nutritional value of halloumi cheese

Halloumi cheese is considered a rich source of calcium (approximetly 700-794mg/100g) which is comparable with Edem and Gouda cheese (795 and 773mg/100mg, respectively). The estimated average requirement of calcium for adults (age between 19 and 50 years old) is 525mg/day (Anonymas, 1995). Moreover, Halloumi cheese is a good source of protein and fat with value around 22g/100g and 26.1g/100g respectively. (Papademas, 2000).

2.6. Milk Coagulating Enzyme Used in Halloumi Cheese Making

Milk clotting enzymes are one of the most significant cheese making raw materials impacting and regulating milk coagulation properties. Rennet is a generic name for an enzyme prepared and used to coagulate milk in the production of rennet-coagulated cheese and rennet casein (Fox and Mc Sweeney, 1997). Milk coagulation properties are of great importance as they significantly influence cheese yield and quality (Kubarsepp *et al.*, 2005). Many different types of rennet and coagulants are, or have been, used for manufacturing of different types of cheeses. The types of rennet and coagulants as well as their characteristics have been reviewed by several authors (Guinee and Wilkinson, 1992; Wigley, 1996). Rennet and other coagulants are most efficiently categorized according to their source.

2.6.1. Animal rennet and coagulants

Within the group of products of animal origin, calf rennet is regarded as the ideal enzyme product for cheese making due to its high content of chymosin, nature's own enzyme for coagulating bovine milk. In the abomasums and extracts from its tissues, the proportion varies between the two enzymes, chymosin and pepsin, depending on the age of the animal and the type of feed (Andr'en, 1982). Extracts from young calves have high proportion of chymosin content, typically 80-95 IMCU (international milk clotting units) 100 $IMCU^{-1}$ chymosin and 5-20 IMCU 100 $IMCU^{-1}$ pepsin. Adult bovine rennet is an extract from mature animals and has a much higher content of pepsin, typically 80-90 IMCU 100 $IMCU^{-}1$. Throughout the world, animals are slaughtered at different ages and all kinds of mixtures of the extracts exist, resulting in a broad range of composition for commercial rennet (Rampilli, 2005).

The traditional product, calf rennet, has until recently been the reference product against which alternative products are measured. Adult bovine rennet is the most widely used alternative to calf rennet, which is not surprising as it contains the same

19

active enzymes. The high pepsin content in adult bovine rennet gives the product a high sensitivity to pH, and a higher general proteolytic activity. Several niche products exist, of this lamb/ovine and kid-caprine/caprine rennet are very similar to calf/adult bovine rennet, but they are best suited for clotting milk of their own species (Foltmann, 1992).

2.6.2. Microbial coagulants

All the well-known microbial coagulants used for cheese making are of fungal origin (Repelius, 1993). Most of the bacterial proteases described as milk-clotting enzymes have been found to be unsuitable, mainly because they have high proteolytic activity. Of the microbial coagulants used for cheese making, *Rhizomucor miehei* is predominant microbial coagulant. It exists in four types, all significantly more proteolytic than chymosin (Fox and Stepaniak, 1993).

2.6.3. Fermentation-produced chymosin (FPC)

FPC is chymosin produced by fermentation of a Genetically Modified Organism (GMO). The product contains chymosin identical to that from animal source, meaning that they have the same amino acid sequence as chymosin from the corresponding animal stomach, but it is just produced by more efficient means. FPC products have been on the market since 1990. The main FPC, which contains bovine chymosin, is today considered to be the ideal milk-clotting enzyme against which all other milk-clotting enzymes are measured. The production and application of bovine-type FPC has been reviewed (Repelius, 1993). Recently, a new generation of FPC, identical to camel chymosin, has been developed. FPC (camelus) has been found to be an even more efficient coagulant for bovine milk than FPC (bovine), and is among others characterized by their very high specificity against caseins, which leads to high cheese yields without creating any bitterness (Harboe, 1999).

2.6.4. Vegetable coagulants

Many enzymes from plants have been found to coagulate milk, but the one extracted from *Cynara cardunculus* (L.) cardoon, seems especially suitable. Since ancient times, the flowers of *C. cardunculus* have been used in artisanal cheese-making, especially in Portugal, where it is considered a superior for cheeses such as Serra and Serpa. Cardoon coagulants are not widely used, but they are produced and used locally in some Mediterranean countries (Garg and Johri, 1994).

2.7. Molecular and Catalytic Aspects of Enzymes in Rennet and Coagulant

The molecular aspects of the milk-clotting enzymes present in rennet and coagulants are important for the understanding of the similarities and differences between the products. All enzymes primarily used for making cheese belong to the family of aspartic proteases, which are characterized by a high content of dicarboxylic and hydroxy amino acids and a low content of basic amino acids. (Chitpinityol and Crabbe, 1998).

The isoelectric point and proteolytic pH optimum of all aspartic proteinases are acidic, although the milk-clotting enzymes have high activity at almost neutral pH (6.5). The general proteolytic pH optimum of chymosin is 3.8, but it has high specific milk-clotting activity at the pH of milk, that is, 6.7. Compared with pepsin, which has its general proteolytic pH optimum at about 2-4, the milk clotting activity of 1 mg chymosin corresponds to that of about 5 mg pepsin at pH 6.7. The catalytic mechanism of the milk-clotting enzymes is to hydrolyze the Phe105–Met106 bond of k-casein on the casein micelle surface. Hydrolysis of k-casein destabilizes the casein micelles, which coagulate in the presence of Ca^{+2} (Harboe *et al.,* 2010).

Chymosin has a strong affinity for this region of k-casein and has the highest specific milk-clotting activity of the aspartic proteinases. The other milk-clotting aspartic proteinase, which is, pepsin, has a specific milk-clotting activity that is lower than that of chymosin. It must also be remembered that milk-clotting conditions, such as pH, calcium content, and temperature, strongly influence milk-clotting activity; the milk-clotting activity of pepsin is especially highly pH dependent. Ovine and caprine chymosins have been shown to have a higher specificity for ewes' and goats' milk, respectively, than for cows' milk (Crabbe, 2004).

2.8. Factors Affecting Rennet Coagulation

2.8.1. Heat treatment

Pasteurization of cheese milk can be attractive for two reasons: (a) control of the microflora by increased reduction of microorganisms in the raw milk and (b) increased cheese yield. The latter effect is attained by denaturation of whey proteins and formation of complexes with the casein leading to a larger part of whey protein being retained in the cheese (Lucey, 2002).

Heat treatment of milk results in a number of changes in physico-chemical properties. These include denaturation of whey proteins: and the interaction between the denatured whey proteins and the casein micelles (Fox *et al*, 2000). As revealed by Horne and Banks (2004), residual-clotting activity in whey protein products is highly unwanted, and it is therefore important that the coagulant can be inactivated by a heat treatment that does not damage whey protein functionality significantly, e.g. 72◦C for 15 s. The temperature also influences the speed of curd formation as all enzymes tend to give faster curd formation at increasing temperature until they approach their optimal temperature (Lucey, 2002).

2.8.2. Temperature

Optimum temperature for the coagulation of milk varies depending on pH and rennet type. The optimum coagulation temperature for the firmness of rennet-induced gels is 30–35 ^{0}C, and the typical coagulation temperature used in cheese-making is 32^{0}C (Lucey, 2002). Prolonged cold storage of milk prior to renneting can result in a longer Rennet clotting time (RCT) and a weaker gel due mainly to dissociation of k-casein from the micelles. This can be reversed to some extent by pasteurization of the milk prior to cheese making (Fox *et al*., 2000).

2.8.3. pH

The pH has a significant effect on the renneting reaction and the properties of the rennet gel. The influence of pH is one of the main process factors affecting the curd formation. The dependency on temperature is also influenced by pH (Horne, 1998). The lower the pH, the better they tolerate higher temperature because the enzymes are more stable at lower pH and calcium phosphate is dissociated from the micelles. pH has also strong influence on the rate of syneresis, and it is important in controlling the amount of whey expelled. Decreasing pH normally increases the syneresis rate of curd (Fox *et al.*, 2000).

Optimum pH for the action of (purified) caseins in milk is 6.3-6.0. Lowering the pH of milk leads to a reduction in the RCT (due to reduced electrostatic repulsion) and a faster rate of increase in gel firmness (Fox *et al.*, 2000). Several factors are involved in this phenomenon, including a reduction in the electrostatic repulsion between micelles, increased $[Ca^+2]$ due to solubilization of colloidal calcium phosphate (CCP), flocculation at a lower degree of k-casein hydrolysis, and increased rennet activity. Acidification of milk increases the strength of rennet-induced milk gels up to pH 6.3–6.0. At lower pH values (i.e., <6.0), gel strength is reduced and there is an increase in the loss tangent probably due to excessive solubilization of CCP, which acts as a cross linking agent between casein molecules and casein micelles. pH dependency differs between the types of rennet coagulants (Horne, 1998).

2.8.4. Rennet concentration

24

The rate of the enzymatic hydrolysis of k-casein in milk is proportional to the amount of rennet added. Dosage of coagulants is mostly calculated in international milk clotting units (IMCU100 L−1) In general, the coagulant dosage is related to the proteolytic specificity with the relation that higher specificity leads to lower IMCU needed to coagulate within a specific time set. Rennet concentration has generally not been found to have any great influence on the properties of the fresh curd or the extent of syneresis in the vat (Lelievre, 1977; Luyten, 1988; Spangler *et al.*, 1991; Kindstedt *et al.*, 1995), though systematic investigations are lacking. Higher rennet concentration results in the enzymatic reaction being faster and running to a higher degree of k-casein hydrolysis before aggregation can reach the same degree as at a lower rennet concentration (Fox *et al.*, 2000).

2.8.5. Calcium chloride

Calcium is often added to the milk during cheese making in the form of CaCl2, leading to increased rates of both the enzymatic and aggregation reactions. Addition of calcium chloride (CaCl2) to cheese milk decreases the pH, reduces clotting time and speeds up the curd formation (Harboe *et al*, 2010). In a typical cheese production site, 0-20g CaCl2 100 kg^{-1}milk is added before the addition of the coagulant, without affecting final cheese quality. This is partly caused by the decrease in pH resulting from CaCl2 addition that affects the rates of both the enzymatic reaction and the aggregation reaction. Addition of CaCl2 to the milk reverses the effect of heat treatment on coagulation time and gel firmness (Lucey *et al.*, 1994; Zoon, 1994) though the coagulation does not proceed in the exact same manner.

2.9. Cutting Time Determination

Curd strength is a general term, arising from a cheese maker's sensory concept of a gel that was sufficiently intact to be cut into discrete grains. For the manufacture of natural cheese, the coagulum (gel), when sufficiently firm, needs to be cut into discrete grains, which expel whey without fragmenting. For this purpose, there is an acceptable range of curd firmness or strength, occurring some time later than the point of gelation, implying a need to measure or infer the firmness of a gel as it forms and up to the point where it is ready to cut (Richardson, 1985).

Coagulum is usually cut in cheese plants after a predetermined enzymatic reaction time has elapsed or upon the operator's judgment of cutting time based on subjective evaluation of textural and visual properties of the curd. Cutting the curd after a predetermined time is questionable because factors that affect curd firmness could cause a variation in the optimum cutting time (Gunasekaran and Ay, 1996). Cutting the curd by relying on the subjective judgment of the operator can be accurate and acceptable if the evaluation is made properly (Hori, 1985). If the gel is too firm at cutting time, syneresis will be retarded, resulting in a high moisture cheese. If the curd is cut too soft then cheese yield will be decreased as a result of increased loss of fat and curd fineness in the whey (Hori, 1985; Payne, Hicks and Shen, 1993).

The cutting of the curd is normally done either at a predefined time after rennet addition or when the cheese maker empirically determines that the curd has the right properties for cutting; often by cutting the curd with a knife and visually evaluating the surfaces and the splitting of the milk gel. In traditional practice, the 'finger' or

26

'knife test' was used to monitor curd strength. This is an empirical assessment, based on experience, in which the cheese maker makes a slight cut in the coagulum with a finger or knife and lifts the curd to see if there is a clean break and if clear whey is exuded, providing an indication of firmness and readiness for cutting. The firmness of the gel at the cutting time varied with conditions. L´opez et al. (1998) measured the gel firmness of the curd at the cutting time, as empirically determined by a cheese maker, in an experiment where rennet concentration and milk pH was varied. Generally, the firmness decreased with decreasing rennet concentration.

3. MATERIALS AND METHODS

3.1. Description of the Study Area

The experiment was conducted in the Dairy Technology Laboratory at Holetta Agricultural Research Center (HARC) of the Ethiopian Institute of Agricultural Research (EIAR). HARC is center of excellence for dairy research at national level. It is located 34 km to the West of Addis Ababa in the central highlands of Ethiopia (2400m above sea level, 38.5^0E longitude and 9.8^0 N latitude). The area has a bimodal rainfall distribution. Over 70% of the rain occurs during the main rainy season (June-September) and about 30% of the rain occurs during the short rainy season (March-May). Forty-three years (1969-2012) meteorological data of the center indicates that the area receives an average total annual rainfall of 1038mm. The long term annual minimum and maximum temperatures of the area were 6.2^0C and 22.4^0C, respectively with mean of 14.3^0C. The major soil type of the area is red brown clay loam nitosol having a pH of 5.1, total N content of 0.2%, p content of 12.4ppm, organic matter (OM) content of 2.2% and cation exchange capacity (CEC) of 17.0 meq/100g soil. (EIAR, 2013).

3.2. Experimental Materials

Fresh abomasums of adult cattle, sheep and goat were collected from one abattoir, and one meat butchers in the vicinity of Holetta. Commercial pepsin powder was used as a control treatment. Different chemicals like sodium chloride, calcium chloride and acetic acid were used to extract the crude enzyme from the abomasums.

A digital pH meter was used to examine rennet pH and cheese pH value. Water bath at temperature of 32 ^{0}c was used to inoculate the samples during the rennet strength and curd determination. Electrical weighing balance was used to take weight of curd as well as of chemicals during analysis. Different culture medias such as plate count agar, potato dextrose agar and violet red bile agar were used during microbiology evaluation.

3.3. Preparation of Liquid Rennet

Rennet was prepared according to the method described by O'Connor (1993). The basic steps of rennet making are shown in Figure 1. After removing the internal contents, abomassal tissues was cleaned with tape water internally while their veins and fat contents were removed externally. Then they were inflated with air and dried in wooden- wire mesh cabinet placed in an open air. After drying, each abomasum was cut into very thin strips separately, weighed and soaked in 1000 ml solution of the 150g NaCl and 40g calcium chloride and 10ml acetic acid solution individually. The solution was stored at 24-25 °c. The extract was filtered with muslin cloth. After filtration a yellowish clarified solution (approximately 950 ml) was obtained.

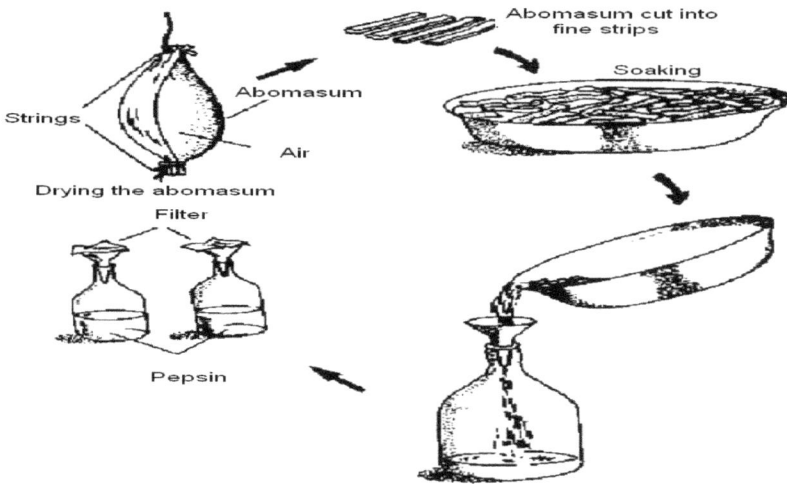

Figure 1. Schematic diagram of rennet making

Source: O'Connor, 1993

3.4. Halloumi Cheese Making

Halloumi cheese was produced by slightly modifying the process line indicated by O'Connor (1993) from cow milk. 80L of cow milk was collected (in different time) from Holetta Agricultural Research Center dairy farm. Milk was pasteurize (heat treated) to 72°C for 15 seconds and cooled immediately with cold water to 32°C. Then the crude enzyme extract was added to the cheese milk. Curd was expected in 40- 45 minutes. Then the curd was cut in to a 3- 4cm cubes with a sharp cutting knife. Stirring of curd and whey was then followed by heating at $38\text{-}42^0c$. The curd was left to settle to decant the whey off. Once the curd was separated from the whey, then the content was scooped in to a mould lined with cheese cloth and pressed for about four hours. After the whey was collected, it was heat treated and the curd removed from the press cut into the desired size and placed in heated whey. After 20 minutes, the curd pieces was transferred to the draining table and allowed to cool. Finally, the cold curd was placed in clean and dried closed dish for quality and microbiology examination.

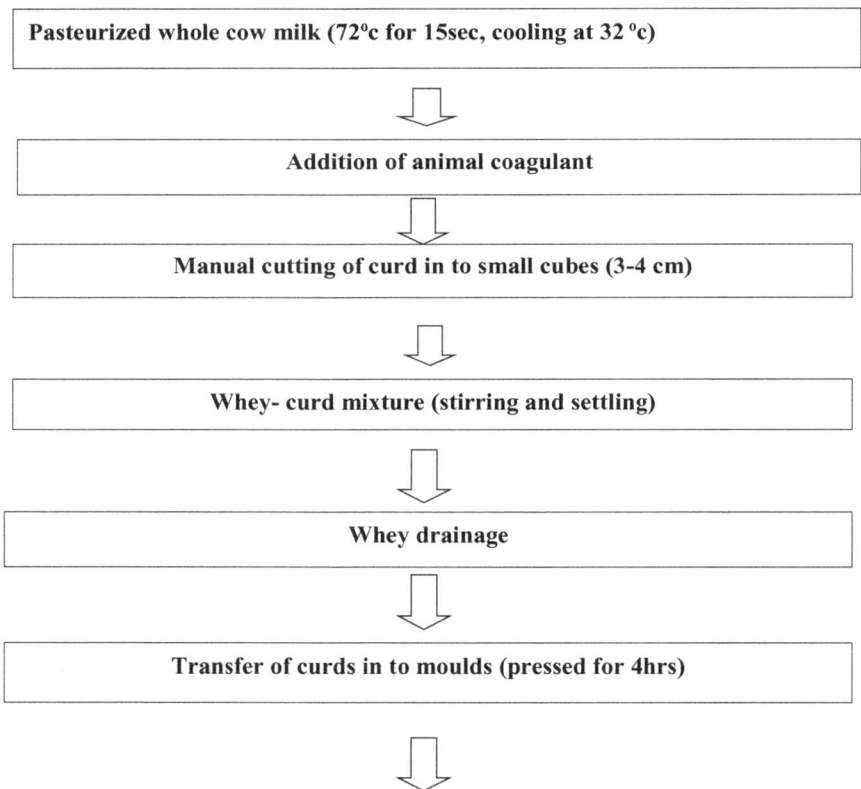

Pasteurized whole cow milk (72°c for 15sec, cooling at 32 °c)

⬇

Addition of animal coagulant

⬇

Manual cutting of curd in to small cubes (3-4 cm)

⬇

Whey- curd mixture (stirring and settling)

⬇

Whey drainage

⬇

Transfer of curds in to moulds (pressed for 4hrs)

⬇

Cooking curd block in hot whey (90 ^{0}c for 20minutes)

⇩

Dry salting of cheese

⇩

Fresh Halloumi Cheese

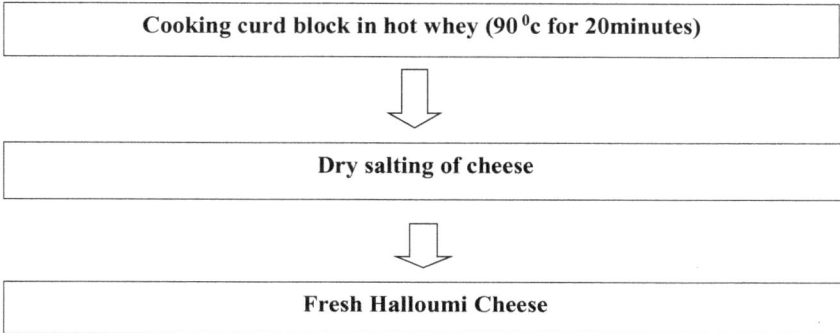

Figure 2. Halloumi cheese making flow chart

Source: O'Connor, 1993

3.5. Experimental Procedure

Rennet was made as described above and kept until it loses its recommended strength of coagulating milk in 40-45 minutes. Then the two rennet types and control was further tested for Halloumi cheese quality and yield.

Preliminary observation

In the current study, the rennet produced was crude extract and the amount of pepsin present in the solution was unknown. It was therefore necessary to establish at least indicative strength of the rennet before applying it to milk and calculate the appropriate dosage that was needed for optimal curdling and the quality of the resulting cheese.

To determine the correct amount of rennet added into the cheese milk, preliminary observation was conducted based on the standard used by O'Connor (1993). Different rennet concentrations were set to identify the minimum curd forming concentration (Table 1).

Table 1. Rennet concentrations set to determine the minimum curd forming concentration

Rennet concentrations tested in ml/liter of milk						
Cattle rennet	Sheep rennet		Goat rennet			
0.3	0.3	10	0.3	10	45	350
0.5	0.5	15	0.5	15	50	400
1	1	20	1	20	100	450
2	2	25	2	25	150	500
3	3	30	3	30	200	
4	4	35	4	35	250	
5	5	40	5	40	300	

Caprine rennet showed extremely weak curd and unable to form cheese with the aforementioned rennet concentrations used. Therefore, caprine rennet was dropped from the present study. Based on the minimum curd forming potential, bovine crude rennet at a concentration of 2ml/lt of milk and ovine crude rennet at a concentration of 30ml/lt of milk were selected and tested.

3.6. Experimental Treatments

The two rennet types extracted from Bovine, and Ovine abomasums was used as treatments of the study. In addition to this commercial pepsin powder was used as a positive control for experimental procedure.

3.7. Experimental Design

The experiment was laid out in completely randomized design (CRD) with two enzyme types (pepsin and commercial rennet) and two species (bovine, and ovine). Three treatments i.e. rennet types with different animal species and four replication for each were used.

3.8. Observations

3.8.1. Rennet strength

Rennet strength is conventionally measured by determining the time required to coagulate milk under standardized condition. The clotting activity (strength) of rennet types was measured and reported in clotting strength according to (Berridge, 1952). In addition, the time length in day's rennet maintains its strength was also determined by testing the extracted rennet samples every week. Rennet pH and curd pH was measured in each evaluation analysis.

Measurement of Rennet clotting time

The method of Berridge (1952) was used to measure clotting time. The main steps were as follows. The standard substrate was prepared from cow milk heated at 32 ^{0}c at 10% (w/v) solution of cacl$_2$ (0.01M) solution. The prepared animal rennet 1ml/10ml of standard substrate was added and mixed manually and incubated in a water bath at 32°C. After thoroughly mixing three times, the zero clotting time started. The milk clotting activity of each extract was measured, with the assumption that all the soluble proteins from the extract were enzymes which coagulate milk at 32°C. The clotting activity equation as reported by Berridge (1952) in rennet units (RU) was used.

RU = 10 x V/Tc x Q

RU: rennet unit
V: volume of standard substrate (ml)
Q: volume of animal rennet (ml)
Tc: time of clotting (sec)

Clotting strength

The clotting activity of rennet types was also reported in clotting strength of Soxhlet (F) based on the equation of (Bourdier and Luquet, 1981).

F = RU/0.0045

35

3.8.2. Chemical analysis

In the experiment, the major chemical composition of Halloumi cheese was evaluated for determination of fat, protein, moisture, total solids, ash, pH and acidity.

Fat content: the fat content in cheese was determined by Gerber method following the procedures outlined by (O'Connor, 1995). Three gram of cheese sample was weighed out in a piece of greaseproof paper. Then 10ml of sulphuric acid was dispensed in to the butyro meter followed by carful addition of water so that it rests on the acid. The cheese sample was then wrapped from cylinder that fits in to the butrometer. Then additional water (4-5 ml) was further added. Amyl alcohol (1ml) was then added in the cheese sample. Stopper the butyro meter securely and was shaken to dissolve the cheese. It was difficult to dissolve the cheese; in this case the butro meter was placed in the heated water bath and removed periodically for mixing until the cheese was fully dissolved. Finally the butro meter was centrifuged and reading was recorded as for milk and cream.

pH.: The pH of each cheese sample was measured following the procedures outlined by O'Connor (1995) through electronic digital pH meter. The electrode was immersed directly into the cheese sample (curd) until the pH sensitive bulb was covered. The temperature of the sample was measured and the pH meter was activated and pH reading was recorded.

Acidity: Acidity was measured following the procedures outlined by O'Connor (1995) by titration (O'Connor, 1995). Ten gram of cheese sample was prepared, 105 ml of water at 40°C was added into the cheese sample and the content was shaken vigorously. 25 ml portion (2.5 g) of the filtrate was titrated with standard 0.1N NaOH solution (until definite pink color persists) using phenolphthalein indicator. The result was expressed as % lactic acid.

$$\text{Lactic acid (\%)} = \left[\frac{\text{ml} \frac{N}{10} \text{alkli} \times 0.009}{\text{ml of sample used}} \right] \times 100$$

Total Solids: 5 grams of Halloumi cheese samples was weighed on a crucible in duplicate and kept in a constant temperature oven set at 105°C for 12 hrs until a constant weight was achieved. The loss in weight was calculated as percent moisture and the total solids was calculated subtracting percent moisture from 100 (Kirk and Sawyer, 1991).

$$\text{Total solid} = \frac{\text{crucible weigth} + \text{oven dry sample weight} - \text{crucible weigth}}{\text{sample weigth}} * 100$$

Ash: A fresh Halloumi cheese sample was ignited at 600°C for 3 hours and weight when cooled to room temperature. Loss in weight was calculated as percent organic matter and the ash content was determined by subtracting the organic matter percentage from 100 (Kirk and Sawyer, 1991).

Protein content: Total protein content of the cheese samples was determined by the Kjeldahl method (AOAC, 1995).

Digestion: 5 g of the cheese sample warmed in a water bath at 38 °C and poured into a Kjeldahl flask. 15g potassium sulphate, 1.0 ml of copper sulphate solution and 25 ml of concentrated sulphuric acid was added into the flask and mixed gently. The digestion was carried out in a digestion block until a clear solution is appeared. Then it allowed to cool at room temperature over a period of 25 minutes.

Distillation: The Kjeldahl tubes were placed in the distillation equipment. Seventy five ml of 40 percent sodium hydroxide solution was added slowly down the neck of the flask. Then ammonia was distilled into 50 ml boric acid solution with bromocresol green/ methyl red indicator until blue color appeared. Finally, the sample were titrated with 0.1N hydrochloric acid solution from a burette until a faint pink color solution is formed and the burette reading was taken to the nearest 0.01 ml. The blank test was carried out using the above procedure by taking 5 ml of distilled water with about 0.85 g of sucrose instead of the test portion. The percentage of nitrogen in the cheese samples was calculated (AOAC, 1995).

$\%N = (1.4007 * (Vs - Vb) * N\ HCl * 100) / \text{Weight of sample}$

  Where:

$\% N$ = Percentage nitrogen by weight

Vs = Volume of HCl used for titration of sample

Vb = Volume of HCl used for titration of the blank

$\%CP$ = Percentage of crude protein

$\%CP = \%N * 6.38$

3.8.3. Cheese quality

Cheese quality was evaluated by determining organoleptic scores (sensory evaluation) based on appearance, odur, and texture and consistency of the cheese samples. Five volunteer experienced cheese consumers were selected. The panelists were mainly researcher and graduate students in the dairy science department of Holetta Agricultural Research Center. Clean water held at room temperature was served for cleaning mouth during testing samples. The judges evaluated the appearance, flavor, texture and consistency of Halloumi cheese. Hedonic scale was used to evaluate the Halloumi cheese in which, 5 refers to excellent, 4 to good, 3 to fair, 2 to poor and 1 for unacceptable quality of Halloumi cheese (Lawless and Heymann, 1999).

3.8.4. Micro biological evaluation

The microbial tests considered were total Plate Count (TPC), yeast and mould (YMC) and coliform count (CC). For determination of total plate count, yeast and mould and coliform count, peptone water was sterilized by autoclaving at 121°C for 15 minutes. Similarly the total plate count agar used for determination of total viable organisms was sterilized by autoclaving at 121°C for 15 minutes; Potato dextrose agar was used for determination of yeast and mould. Violet red bile agar (VRBA: Oxoid) was used for determination of coliform count and was sterilized by boiling (Richardson, 1985). For the above tests, the media used were prepared according to the guidelines given by the manufacturers.

Total plate count The total bacterial count was made on plate count agar (PCA) by adding 1 g of cheese sample into sterile test tube having 9 ml peptone water. After thoroughly mixing, the sample was serially diluted up to $1:10^{-7}$ and duplicate samples (1 ml) were pour plated using 15-20 ml standard plate count agar solution and mixed thoroughly. The plated sample was allowed to solidify and then incubated at 32°C for 48 hours. Colony counts were made using colony counter (Marth, 1978).

Yeast and Mould Count was also be made by appropriate decimal dilutions of cheese pour-plated in duplicate on potato dextrose Agar and colonies were counted after incubating the plates at 25°C for 5 days (Luck and Gavron, 1990).

Coliform count (CC): One ml of cheese sample was added into sterile test tube having 9 ml peptone water. After mixing, the sample was serially diluted up to $1:10^{-5}$ and duplicate samples (1 ml) were pour plated using 15-20 ml Violet Red Bile Agar solution (VRBA). After thoroughly mixing, the plated sample was allowed to solidify and then incubated at 30°C for 24 hours. Finally, colony counts were made using colony counter (Marth 1978). Typical dark red colonies were considered as coliform colonies.

Average numbers of microorganisms was used to calculate the total colony forming units by multiplying by the dilution factor to get the colony forming units per gm and this was then transformed to its natural logarithmic value.

3.9. Total Cost of Production

The total cost of locally produced rennet was determined by comparing and calculating cost incurred to produce 1 kg of cheese. Positive economic balance (minimum coagulant cost and maximum cheese yield) was used as a distinction mark for alternative rennet substitute.

3.10. Statistical Analysis

All experiments were performed with four replicates each. All the data were reported as means with standard deviations. An analysis of variance (ANOVA) was applied to assess differences among the "Gastric Enzyme Extract from adult Bovine and ovine" and the commercial enzymes by using SAS Software version 9.1. A repeated measure analysis was applied to assess storage condition (rennet pH and curd pH) of liquid rennet. Failure data analysis was run to fit the expiry date data to Weibull survival distribution to model the shelf-life data (rennet) as used by (Schmidt and Bouma, 1992).

Data for nutritional contents were subjected to analysis of variance (ANOVA) to test for significant differences at $P<0.05$. All microbiological data were transformed into Log10 CFU/g before subjecting to statistical analysis. For each experiment mean comparisons were done separately using least significant difference (LSD) test for variables whose F-values showed significant difference at 5 % significance level.

Models

1. Observation Model: for rennet strength and nutritional content of Halloumi cheese were:

$Y_{ik} = \mu + \alpha_i + \varepsilon_{ik}$ i= treatment levels (1, 2, 3)

Where,

Y = processing parameters, quality parameters

μ = Overall mean

α = effect of treatments

ε = random error, which has normal distribution with mean zero and a constant variance

2. The model used for repeated measure of stored rennet was:

$Y_{ijk} = \mu + \alpha_i + \delta_{ij} + \beta_k + \varepsilon_{ik}$

Where,

μ = Overall mean

α_i = effect of coagulant enzyme source (bovine, ovine, commercial rennet)

β_k = effect of periods (weeks)

δ_{ij} = random error with mean 0 and variance δ^2s. The variance between replications (subject) with in treatment and it is equal to the co variance between repeated measurements within treatments.

ε_{ijk} = random error, which has normal distribution with mean zero and a constant variance

3. Weibull model: The model used for failure analysis was Weibull distribution with the general function;

$$f(x; k, \lambda) = \frac{k}{\lambda} \left(\frac{x}{\lambda}\right)^{k-1} e^{-(x/\lambda)^k}$$

; $x > 0$. (x = failure day)

$k > 0$ is the *shape parameter* and $\lambda > 0$ is the scale parameter of the distribution. $k < 1$ indicates that the failure rate decreases over time. If the failure rate is constant over time, then $k = 1$. $k>1$ indicates the failure rate increases over time

4. RESULTS AND DISSCUSSION

The results obtained in the current study are presented in three sections: clotting ability and shelf life of rennet extracts from three different sources; yield, chemical composition, and microbial and sensorial qualities of Halloumi cheese produced using three different milk coagulants; and production cost of rennet extracted from mature ovine and bovine abomasums, discussions are also made under the respective sub titles.

4.1. Characterization of Rennet Strength and Shelf Life

4.1.1. Clotting activity

At individual observation level, the clotting activity of the milk coagulants used ranged from the lowest 0.0074RU for ovine pepsin to the highest 0.196RU for commercial pepsin (data not presented) with the overall average value being 0.13RU (Table 2). On average, the highest clotting activity (0.1895 RU) was observed for the commercial pepsin, while the lowest value (0.023 RU) was recorded for ovine pepsin (Table 2). No apparent (P>0.05) difference was observed between commercial and bovine pepsins in their clotting activity, while ovine pepsin showed significantly (P<0.05) lower clotting activity compared with the other two (Appendix, table 1 and 2). Of the two locally produced pepsins, bovine pepsin had clotting activity 0.1565 RU higher (about 780%) than that of ovine pepsin (Table 2). It can also be observed that average clotting activity increased with increasing Soxhlet strength and decreasing clotting time. Accordingly, the highest clotting activity recorded for the commercial pepsin corresponded with the highest Soxhlet strength and shortest clotting time.

Variation in milk clotting activity (MCA) and strength might be related to differences in the nature of the clotting ability of raw materials, the proteolytic efficiency and/or the concentration of the rennet types on the milk substrate used. For instance, as indicated by Guinee and Wilkinson (1992), the MCA of rennet relies on its ability to degrade casein micelles, the action being dependent on the chymosin and pepsin content of the rennet complex. As a critical parameter in the progress of curd formation, clotting time can be usefully related to curd firmness and curd yield (Ramet, 1997). The weak clotting ability of ovine rennet observed in the present study might be attributed to the weak such property of pepsin present in the rennet solution. This observation is in agreement with that of Vairo Cavalli *et al.* (2005), who indicated that the breakdown of cow milk by ovine caseinate took place much slower than that by bovine counterpart. The weak clotting power of ovine pepsin on cow milk might be related to species specificity. Calvo and Fontecha (2004) for instance, indicated that the clotting ability of each rennet proteolyses is stronger for their species specific casein compared with that of other species. On the other hand, if milk takes a short time to coagulate, it leaves more time for curd firming and obviously has better coagulation ability in general; thus, the final curd will be firmer. Conversely, if milk takes a long time to coagulate, the curd will have less time to get firm thus will be smoother (Fuquay *et al.,* 2010).

Table 2. Mean (± SD) clotting time, change in clotting activity (rennet unit: RU) and strength of Soxhelt (F) for the three milk coagulants tested

Milk Coagulants	No. of observations	Clotting time (min)	Clotting activity (RU)	Strength of Soxhelt (F)
Overall Mean	12	13.43	0.13 ±0.14	33.3± 1.57
Bovine pepsin	4	9.5[a]	0.1795[a]±0.02	39.97[a] ± 2.42
Ovine pepsin	4	22[b]	0.0230[b]±0.02	16.92[b] ± 2.42
Commercial pepsin (C)	4	8.8[a]	0.1895[a]±0.02	42.20[a] ± 2.42

Means within the column followed by different superscripts are significantly different (P ≤ 0.05)

4.1.2. Stability of rennet from bovine, ovine and commercial source

The stability of all rennet sources were not affected during the 6 weeks of storage i.e. the failure probability (hazard) of losing stability was zero (Figure 3). The commercial pepsin didn't show any stability failure (hazard) over the ten weeks storage period. This might be due to the short storage period set for the present experiment. Between the locally produced pepsin, ovine pepsin started to lose its stability after six weeks of storage, while, bovine pepsin remained active for extra three weeks (till the 9th weeks of storage) (Figure 3). The changes that have taken place during the storage period might be resulted from the manufacturing process, the storage temperature as well as the extent at which the rennet were exposured to light. For instance, as indicated by Arvanitoyannis (2009), rennet, when dried, can resist to even very high temperature while in a related study, Pometto et al. (2006), revealed that liquid rennet solutions were less stable and the substance decomposed at various speeds. In the current study, the rennet (pepsin) were stored in a dark closed container and as reported in earlier related works, this may help the enzymes to keep their stability (Arvanitoyannis, 2009). However, the ambient room temperature at which the rennet was stored may affect their stability by creating heat variation. Such a condition was reported by Teply et al. (1978) who indicated a reduced liquid rennet efficiency kept at room temperature with the intensity of efficiency reduction being a function of the strength of the solution. Arvanitoyannis (2009) also stated that, the rennet strength decreases when the substance is handled and stored improperly, with light, heat and shaking exerting clear detrimental effects on the rennet.

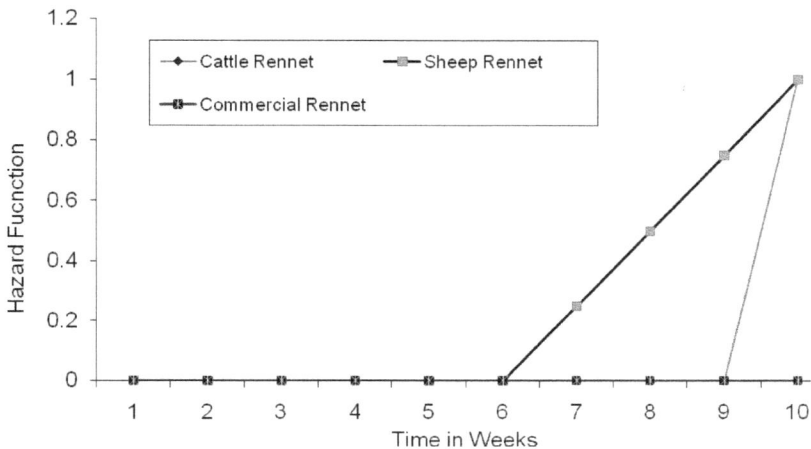

Figure 3. Cumulative failure plots of rennet extracted from cattle, sheep and commercial rennet

The pH value of stored rennet ranged from 4.21 to 3.45 (Table 3). The repeated measure analysis revealed that highly significant difference ($P< 0.05$) observed between pH values of locally produced and commercial rennet used in the current study (Appendix Table 4). The commercial rennet (control) showed no pH variation over the ten-week storage period. Least Significant Difference (LSD) comparison of means at rejection level of 0.05 revealed that means were significantly different from each other ($P< 0.05$).

Table 3. Mean (± SD) rennet pH and curd pH evolution of stored rennet for the three milk coagulants tested

Treatments	No. of observation	Rennet pH	Curd pH
Over all mean	12	3.70	5.8
Bovine pepsin	4	$3.45^a \pm 0.006$	$5.94^a \pm 0.007$
Ovine pepsin	4	$3.46^a \pm 0.006$	$5.34^b \pm 0.005$
Commercial pepsin	4	$4.21^b \pm 0.006$	$6.12^c \pm 0.007$

Means within the column followed by different superscripts are significantly different ($P \le 0.05$)

Over the storage period, the pH values of the locally produced liquid rennet (pepsin) tended to continuously decline (Figure 4). Davis (1965), noted that pH plays a significant role in the stability of the coagulant enzyme. The variation in rennet pH during storage might be caused by one or a combination of lack of standardized manufacturing protocols, purification methods and storage temperature used in the processing. As reported by De Caro et al. (1995), a substantial loss of rennet could also occur due to inefficient rennet manufacturing and purification processes. As indicated earlier, crude abomasal rennet extracts were used for the present study and stored at room temperature, and as stated by Hooydonk and Van Den Berg (1988), the strength of natural rennet is not maintained at constant level specially when stored at ambient temperature. The average activity loss per a given unit of time of natural source rennet is relatively high compared with that of the commercial ones. When rennet is kept at high temperatures, pepsin activity increases in whey proteins (Hooydonk and Van Den Berg, 1988).

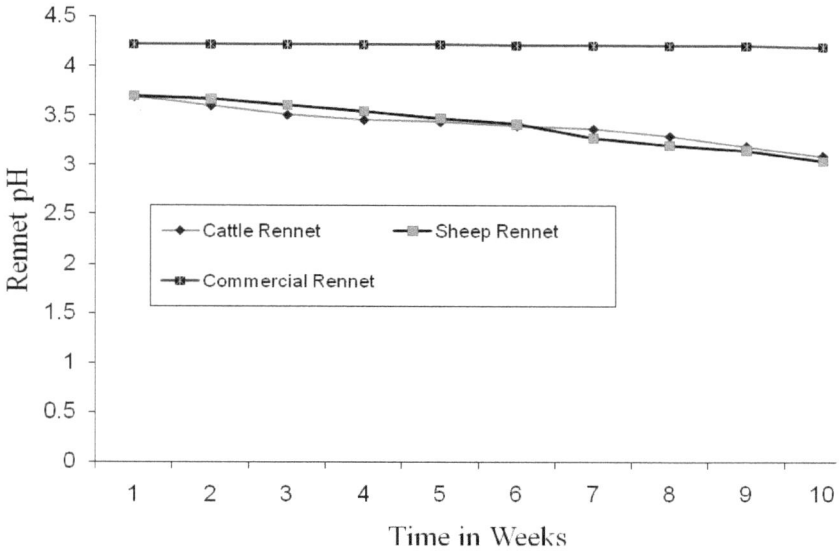

Figure 4. pH evolution of stored mature cattle and sheep abomasal, and commercial rennet

The pH values of the cheese curds formed by the three rennet types at various storage periods (weeks) ranged from 5.34 to 6.12 (Table 3). Milci *et al.* (2005) also reported similar average pH values of Halloumi cheeses made from bovine milk (5.37 - 6.45). Mean curd pH value of repeated measure analysis revealed that highly significant difference (P<0.05) was observed between Bovine, Ovine and commercial rennet

(Appendix table 5). Though rennet storage time advances and cheese curd pH declines, curd formation continued till end of the experimental period. However, the curd pH declining tendency as rennet storage time advances indicates rennet stability losses.

4.2. Effect of Locally Produced Rennet's on Halloumi Cheese Yield and Quality

4.2.1. Chemical composition of milk

The fat, protein and pH values of the cow milk used for the manufacturing of Halloumi cheese were 4.28, 3.25 and 6.22, respectively. Fat and protein are the two primary milk components that are recovered in a cheese-making process and are directly related to cheese yield (Fox *et al.*, 2000). These constituents influence the casein to fat ratio, total solids, lactose and mineral contents, consequently determining the moisture content as well as acid development in the finished product cheese (Traordinary, 2001). However, the casein fraction of the milk protein is the dominant factor affecting curd firmness, syneresis rate and moisture retention ultimately affecting the cheese quality and yield (Lawrence, 1993).

4.2.2. Chemical composition and yield of halloumi cheese

The composition of cheese has a marked influence on all aspects of quality, including sensory properties, texture, and cooking properties (Tunick *et al.*, 2007).

The yield of Halloumi cheese samples ranged from 124.5 to 134.1g with the average yield being 128.4g per liter of milk (Table 4). Treatment had a significant effect on the yield of Halloumi cheese (P≤0.05) (Appendix Table 6). Halloumi cheese made using fresh bovine rennet (pepsin) had higher yield compared with that made using ovine and commercial rennet (pepsin) (Table 3). This observation is in agreement with that of Ahemed *et al.* (2013), who reported higher cheese yield (32.2%) using laboratory produced fresh rennet compared with that produced using commercial rennet (29.8%). Differences in cheese yield using different types of coagulants are likely linked to their proteolytic specificity, as highly specific coagulants produce higher cheese yield. This has been substantiated in numerous cheese yield trials and quality studies comparing various rennet and coagulant types (Emmons *et al.*, 1990; Emmons and Binns, 1991; Banks, 1992; Guinee and Wilkinson, 1992; Quade and Rudiger, 1998). In the current study, bovine and commercial rennet (pepsin) showed higher clotting and short coagulation time (Table 2). The lower cheese yield obtained using ovine rennet might be due to its weak clotting power and non specificity of the rennet to bovine milk. As reported by Kubarsepp *et al.* (2005), milk with favorable coagulation conditions (short coagulation and curd firming times and firm curd) is expected to give more cheese yield.

The moisture content of Halloumi cheese samples ranged from 46.77 to 51.62% with an overall mean value of 49.34% (Table 4). Based on the ANOVA results, no significant differences ($P \leq 0.05$) were observed between cheeses made using bovine and commercial rennet with respect to moisture attributes. These values are in agreement with that reported by Pezeshki *et al.* (2011), where no significant difference was observed in moisture content among cheeses made using differently prepared *Withania* coagulant. Raphaelides *et al.* (2006), on the other hand, reported mean moisture content of 47.40% for Halloumi cheese samples manufactured from bovine milk. With regard to cheese made using ovine rennet, the large amount of rennet (30ml/1Lt) added to the cheese milk reduced the moisture content to significant level (Appendix Table 12). This was because of the large amount of enzyme extracting salts which tend to synerse due to acidity and helped in squeeze out off the moisture. This observation is in agreement with that of Guven *et al.* (2008) who indicated that the moisture content of cheese decreases as rennet concentration increases. The value of the moisture content of Halloumi cheese observed in the present study suggests that it is an intermediate moisture cheese and is much lower than the corresponding value (79%) reported for the traditional Ethiopian cottage cheese, Ayib (Ashenafi, 1992).

The pH value of Halloumi cheese ranged from 5.72 to 6.16 with an average pH value of 6.0 (Table 4). On average, the highest pH value (6.16) was observed for commercially made Halloumi cheese. No apparent ($P>0.05$) difference was observed between Halloumi cheese made using commercial and bovine pepsins in their pH value, while Halloumi cheese made using Ovine rennet showed significantly lower pH value (Appendix Table 7). The acidic features of the rennet and the minimum requirements of ovine rennet to form curd (30ml/l) could be a factor for this significant difference. For instance, as indicated by Hynes *et al.* (2001), low pH causes the protein matrix in the curd to contract and squeeze out moisture and finally affect cheese yield. According to the findings of Milci *et al.* (2005) and Raphaelides *et al.* (2006), the average pH values of Halloumi cheese samples made from bovine milk ranged from 5.37-6.45. The pH values of Halloumi cheese samples observed in the current study also fall within this range.

The average titrable acidity of Halloumi cheese ranged from 0.049 to 0.027 and was markedly affected by treatment ($P<0.05$). The LSD test showed that the two locally produced rennet types i.e. ovine and bovine rennet (pepsin) differed significantly

(Table 4) (Appendix Table 8). The corresponding pH values and rennet concentrations are the likely attributing factors for this significant difference. In the current study, ovine and bovine rennet differed in their minimum curd forming concentration. The high concentration of ovine rennet used could be related to the significantly lower pH and higher acidity of the cheese samples. The titrable acidity (0.034%) obtained in this study is slightly lower than that reported by earlier studies (0.127%) (Davis *et al.*, 1993).

Table 4. Yield, pH, acidity and proximate composition of Halloumi cheese samples

Constituents	Overall mean	Treatments		
		Bovine rennet	Ovine rennet	Control
No. of observations	12	4	4	4
Yield (g/liter of milk)	128.4	134.1 ± 5.68^a	124.5 ± 5.68^b	126.6 ± 5.68^{ab}
pH	6.0	6.12 ± 0.07^a	5.72 ± 0.07^b	6.16 ± 0.07^a
Acidity, %	0.034	0.027 ± 0.01^a	$0.049\ 0.01^b$	0.026 ± 0.01^a
Total solid, %	46.7	48.26 ± 1.71	45.07 ± 1.71	46.77 ± 1.71
Fat, %	22.5	23.75 ± 4.24	21.25 ± 4.24	22.50 ± 4.24
Protein, %	19.13	19.38 ± 1.12	18.77 ± 1.12	19.26 ± 1.12
Moisture, %	49.34	51.62 ± 1.71^a	46.77 ± 1.71^b	49.62 ± 1.71^a
Ash, %	5.06	5.13 ± 1.49	5.05 ± 1.49	5.01 ± 1.49

Means within the same row followed by different letters are significantly different at $P \leq 0.05$.

The fat content of the Halloumi cheese samples ranged from 21.25 to 23.75% and was not affected (P> 0.05) by rennet type (Table 4) (Appendix table 9). A similar result was also reported by Jeremiah *et al.*, (2004). However, lower values of 5.33% were reported by Frazier and Westhoff (1988). Fasakin and Unokiwedi (1992), on the other hand, reported a much higher value (44.5%) and Milci *et al.* (2005) reported 25.43% for similar cheese samples.

The protein content of Halloumi cheese samples was not affected by rennet type (P>0.05) (Appendix Table 10) though ranged from 18.77 to 19.38% (Table 4). All the mean values observed in the current study are slightly lower than that (19.13%) reported for a similar product by (Milci *et al.*, 2005). However, all the current mean values are higher than the 15% reported for the acid-heat treated Ethiopian traditional cottage type cheese (Ayib) made from bovine milk (O'Mahony, 1988; Ashenafi, 1992; Kassa, 2008).

Total solids and ash contents of the cheese samples were not affected by rennet type (P>0.05) (Table 4) (Appendix Table 11 and 13). The mean total solids values ranged from 45.07 to 48.26%, while mean ash contents ranged between 5.01 and 5.1%. A higher average ash value of 6.52% was reported for Halloumi cheese produced from bovine, ovine and caprine milk (Milci *et al.*, 2005). Kassa (2008) also reported a much lower value (1.16%) for *Ayib* samples. The fairly high ash content of the Halloumi cheese observed in the current study indicates that the product could be considered as a good source of minerals in human nutrition.

4.3. Sensory and Microbial Quality of Halloumi Cheese

4.3.1. Sensory quality

All the cheese samples tested were generally acceptable by the sensory evaluation team (score >3.6). Treatments of cheese with the three different rennet types showed no significant variation on sensory attributes (P>0.05) except for taste (Table 5).

The overall mean scores given by panelists to the appearance, odor, taste and consistency of Halloumi cheese samples were 4.21, 3.96, 3.93 and 4.05, respectively (Table 5). Halloumi cheese produced using bovine rennet received apparently higher (P<0.05) score for taste (4.2), while the lowest score (3.65) was given to Halloumi cheese produced using ovine rennet. Though not significantly different (P>0.05), Halloumi cheese produced using bovine rennet also received higher taste score compared with that produced using commercial rennet. As presented in Table 4, Halloumi cheese produced using ovine rennet received significantly (P<0.05) lower taste score (Appendix Table 19), which might be attributed to the high amount of ovine rennet (30ml/lt) added that in turn might have given the cheese a characteristic of acidic taste.

The non marked difference observed in appearance, odor and consistency of Halloumi cheeses manufactured using locally produced coagulants in the current study was also reported earlier. Gaborit *et al.* (2001) and Martinez-Cuesta *et al.* (2001) for instance, reported that cheeses elaborated with animal rennet and those elaborated with powdered vegetable coagulant had no significant differences (*P* >

0.05) in odor and consistency. Several factors could be attributable to the absence of marked differences in most sensory attributes of such cheeses. As indicated by Hough *et al.* (1999) for instance, variation in sensory perception of panelists could contribute to such situation as different individuals perceive a given food product differently. Generally, as observed in the current study, Halloumi cheese produced from cow milk using locally produced rennet had acceptable sensory quality.

Table 5. Sensory mean scores of cheese treated with different rennet extract

Treatments	No .of observation	Appearance	Odor	Taste	Consistency
Overall mean	12	4.21	3.96	3.93	4.05
Bovine rennet	4	4.1 ± 0.56	4.0±0.31	4.2±0.37[a]	4.15 ± 0.53
Ovine rennet	4	4.35 ±0.56	4.1 ± 0.31	3.65±0.37[b]	3.95 ± 0.53
Commercial rennet	4	4.2 ± 0.56	3.8 ± 0.31	3.95±0.37[a]	4.05 ± 0.53

Means within the same column followed by the same letters are not significantly different at $P \leq 0.05$.

.

4.3.2. Microbial properties of halloumi cheese

Microbial composition of any food product determines its quality, shelf life and safety from food borne illnesses. The microbial profile of cheese is a primary determinant of cheese quality. Microorganisms can contribute to aroma and taste defects.

The total bacterial count of Halloumi cheese ranged from 5.57 to 5.69 $\log_{10}$ cfu/g (Table 6). The mean total bacterial count was significantly different (P<0.05) between Halloumi cheese produced using locally produced rennet and commercial rennet (Appendix table 14). Halloumi cheese produced using ovine rennet had relatively lower total bacterial count (5.57 $\log_{10}$ cfu/g) compared with that produced using bovine and commercial rennet (Table 6). The lower total bacterial count of Halloumi cheese made using sheep rennet might be associated to its lower pH and low moisture content. The high salt content might also have an inhibitory effect on the general micro flora of cheese made using sheep rennet.

The Halloumi cheese samples had yeast and mould counts ranging from 3.87 to 4.05 $\log_{10}$ cfu/g. The mean yeast and mould count differed significantly (P<0.05) (Table 6) between cheeses produced using locally produced rennet, while no apparent (P>0.05) difference was observed between Halloumi cheese made using commercial and bovine pepsins. The markedly higher yeast and mould count observed in cheese made using ovine rennet might be due the higher acidity of the same cheese that creates conducive environment for their growth. Hamed *et al*. (1992), for instance, indicated that yeasts and molds are acid-tolerant and can thus grow on the surface of most cheeses.

When present in milk, coliforms are regarded as "indicators" of post-pasteurization contamination as a result of poor sanitation (Jayarao *et al.*, 2004). In the present study, coliform were detected in Halloumi cheese samples in numbers that ranged from 2.51 to 2.59 log 10 cfu/g (Table 6). The use of different rennet on halloumi cheese did not have any significant (P>0.05) effect on coliform count (Appendix Table 16). The coliform count observed in this study is within the acceptable limit set for pasteurized cheese (10^2/g) and lower than the 3.73 log cfu/g reported for Orgu cheese produced in turkey (Turkoglu, 2003). Ashenafi (1990) and Kassa (2008) also reported higher coliform counts for *Ayib* samples. Such differences might relate to post pasteurization contamination during curd cutting, salting, whey expulsion and moulding. The existence of coliform bacteria in high proportion is suggestive of unsanitary condition or practices during processing or storage (Richardson, 1985). In general, Halloumi cheese produced using locally produced rennet had an acceptable microbial quality.

Table 6. Microbial quality of Halloumi cheese made from Bovine, ovine and commercial rennet (pepsin) source.

Treatment	No.of observation	TBC	YMC	CC
Over all mean	12	5.64	3.94	2.56
Bovine pepsin	4	5.68[a] ± 0.13	3.90[a] ± 0.29	2.59 ± 0.3
Ovine pepsin	4	5.57[b] ± 0.13	4.05[b] ± 0.29	2.59 ± 0.3

| Commercial pepsin | 4 | $5.65^a \pm 0.13$ | $3.87^a \pm 0.29$ | 2.51 ± 0.3 |

Means within the same column followed by the same letters are not significantly different at $P \leq 0.05$.

4.4. Total Cost of Production of Locally Made Rennet and Commercial Rennet

With the intention of cost benefit analysis, a cost of production comparison was made between locally produced rennet and commercial rennet to produce a similar cheese– Halloumi. The two parameters considered were cost of coagulant and cheese yield.

The total cost needed to produce one liter of Bovine and Ovine rennet was 197.20 and 202.2 ETB, respectively, while the cost of commercial rennet was 2500 ETB per 250 ml of rennet (Table 7).

As observed in the present study, 134.1, 124.4 and 126.6g of cheese per liter of milk were produced using Bovine, Ovine and commercial rennet, respectively (Table 7). Based on these figures, rennet production cost were 2.94, 48.76 and 23.6 birr respective for bovine, ovine and commercial rennet a kilogram of Halloumi cheese produced (Table 7). This means that, bovine rennet had (10.5 %) higher cheese yield with lower (80.2 %) production cost followed by commercial rennet. This result provides good evidence that bovine rennet that can be produced from locally available raw materials at lower production cost can be considered as a substitute for the more expensive commercial rennet

61

Table 7. Cost of production of rennet produced from cattle, sheep and commercial rennet source

Input	Bovine rennet (A)	Ovine rennet (B)	Commercial Rennet(C)	A Vs C	B Vs C
Abomasum	40Birr.	45Birr.	-	-	-
Nacl	135Birr.	135Birr.	-	-	-
Cacl$_2$	20 Birr.	20 Birr.	-	-	-
Acetic acid	2.2 Birr.	2.2 Birr.	-	-	-
Total	197.2ETB	202.2ETB	2500ETB/250 ml	-	-
Cheese yield/ L	134.1g	124.4g	126.6g	-	-
Total ml of rennet / kg Cheese	2.94ETB.	48.7 ETB.	23.6 ETB.	+20.6 ETB.	25.1 ETB
Total cost of production	-	-	-	Positive	Negative

5. CONCLUSION AND RECOMMENDATIONS

5.1. Conclusion

In the present study, options of rennet making from adult bovine and ovine abomasum; suitability of locally made rennet for the production of Halloumi cheese; and production cost of the different rennet types considered were examined.

The locally produced bovine rennet demonstrated higher clotting activity, shorter clotting time and higher soxhelt strength, with no marked difference with that of the commercial rennet with all the three parameters considered. The results obtained in the current study provide a better understanding of milk coagulants for Halloumi cheese-making that can be produced from locally available raw materials. Though the locally produced rennet had relatively weaker storage stability compared with the commercial rennet; considering the promising results observed in terms of milk clotting time, milk clotting activity and soxhlet strength: bovine rennet can be considered as an alternative milk coagulant for Halloumi cheese-making, In addition to the aforementioned merits, adult bovine abomasal rennet is locally available with low production cost. Keeping the locally manufactured rennet in a closed container at a dark place and at a temperature below 10°C can be considered to extend its storage stability.

The approximate gross chemical composition of Halloumi cheese produced using locally available coagulants didn't differ markedly with that produced with commercial rennet. The microbial counts considered were also not far from acceptable limits. As judged by sensory evaluation panelists, Halloumi cheeses produced using both locally produced bovine and ovine rennet, were found to be acceptable with respect to overall consumer sensory attributes.

In conclusion, in areas where commercial rennet is not available and expensive, adult bovine rennet could be considered as alternative milk coagulant source for Halloumi cheese-making with regard to cost effectiveness, better clotting strength, easy availability and better cheese yield. The production of bovine rennet in the country can also create employment opportunity and can minimize national expenditure of foreign currency. The adult bovine abomasa rennet and its extraction method described in the current study could be adopted and used by small-scale milk processors.

5.2. Recommendations

- Further research work should be needed to identify the effect of caprine rennet on caprine milk.

- Additional researches should be conducted to further identify the microbial as well as sensory qualities of locally produced rennet and their active ingredients so that better exploitation of them could be possible.

- Further research work is also needed to purify the crud extracts and characterize those using appropriate and recent technologies such as electrophoresis.

- The shelf stability of Halloumi cheese manufactured using locally produced rennet types need also be explored.

6. REFERENCES

Ahmed, G., S.A. Khan, M. Khaskheli, M.A. Qureshi and I. Ahmad, 2013. Production and properties of rennet from buffalo calves abomasam, *Journal of Animal and Plant Sciences,* 23(1 Suppl.): Page: 5-9.

Ahmed Mohamed, Simeon. Ehui, and Yemesrach Assefa, 2004. "Dairy development in Ethiopia." EPTD Discussion Paper. No. 123. IFPRI, Washington, D.C., USA.

Ahmed, Sh., B. Mohamed, P. Hegede and Bekele Tafesse, 2003. Traditional processing of camel meat and milk, and marketing of camels, milk and hides in After Zone of Somali National Regional State, Ethiopia. In Proceeding 10th Annual conference of the Ethiopian Society of Animal Production (ESAP) held in Addis Ababa, Ethiopia. August 22-24, 2002. ESAP, Addis Ababa. pp201-209.

Andr´en, A., 1982. Chymosin and Pepsin in bovine abomasal mucosa studied by use of immunological methods, PhD Thesis, Department of Animal Husbandry, Swedish University of Agricultural Sciences, Uppsala, Sweden.

Anonymous, S., 1995. Dietary reference values for food energy and nutrients for the United Kingdom, No. 41, pp. 136–145. Department of Health, London.

AOAC, 1995. Official methods of analysis, 16th edition chapter 33 dairy products. Association of Official Analytical Chemists, Washington, DC.

Arvanitoyannis, I.S., 2009; Haccp and Iso 22000. Application to foods of animal origin, Blackwell Publishing Ltd, United Kingdom, 560 P. Isbn 9781444320923.

Azage Tegegne and Alemu G/Wold, 1998. Prospects for peri-urban dairy development in Ethiopia. In: proceedings of the Fifth Annual Conference of the Ethiopian Society of Animal Production. 15-17 May 1997. Addis Ababa, Ethiopia.

Banks, J M., 1992. Cheddar-type cheeses. In Encyclopedia of Dairy Sciences, Vol. 1, pp 356–363. Roginski H, Fuquay J W and Fox P F, eds. London: Academic Press.

Berridge, N. J., 1952. An improved method of observing the clotting of milk containing rennin. *Journal of Dairy Research*, 9, 328–329.

Bintsis, T. and P. Papademas, 2002. Microbiological quality of white-brined cheeses: a review. *International Journal of Dairy Technology*, **55**, 113–121.

Bintsis, T., E. Litopoulou-Tzanetaki and R.K. Robinson, 2000. Existing and potential applications of ultraviolet light in the food industry–a critical review. *Journal of the Science of Food and Agriculture*, **80**, 1–9.

Binyam Kassa, 2008. Cottage cheese production in Shashemene and the role of rue (*Ruta chalepensis*) and garlic (*Allium sativum*) on its quality and shelflife. Hawassa, Ethiopia: Hawassa University, M.Sc. thesis.

Brennand, C.P., J.K. Ha, and R.C. Linsay, 1989. Aroma properties and thresholds of some branchedchain and other minor volatile fatty acids occurring in milk fat and meat lipids. *Journal of Sensory Studies*, 4, 105–120.

Bourdier, J.F. and F.M. Luquet, 1981. Dictionnaire laitier. Tec. Doc. Lavoisier, Paris.

Calvo, M.V. and J. Fontecha, 2004. Purification and characterization of a pregastric esterase from a hygienized kid rennet paste. *Journal of. Dairy Science.* 87: 1132-1142.

Castillo, M., 2001. Cutting time prediction during cheese making by near infrared light backscattering. Ph.D. Thesis, University of Murcia.

Chamberlain, A., 1990. An introduction to animal husbandry in the tropics. 4th Edition. Jhon Wiley and Sons INC., New York. 758 pp.

Chitpinityol, M.J. and C. Crabbe, 1998. *Journal: Food Chemistry* - Food Chem, Vol. 61, No. 4, pp. 395-418.

Crabbe, C., 2004. Rennets: general and molecular aspects. In Cheese: Chemistry, Physics and Microbiology, Vol. 1, pp. 19–46. Fox P F, McSweeney P L H, Cogan T MandGuinee T P, eds. Amsterdam: Elsevier.

CSA (Central Statistics Agency), 2013. Agricultural sample survey volume II. Report on livestock and livestock characteristics (private peasant holidings). Addis Ababa, Ethiopia.

Davis, J.G., 1965. Cheese - basic technology. J. and A. Churchill Ltd. London, 4635.

Davis, P.A., J.F. Platon, M.F. Gershwin, G.M. Halpern, C.L. Keen, D. Dipaolo, J. Alexander and V.A. Ziboh, 1993. A Linoleate-enriched cheese reduce lowdensity lipoprotein in moderately hypercholesterolemic adults. Ann. Intern. Med., 119: 555-559.

De Caro, J.; F. Ferrato; R. Verger, and A. De Caro, 1995. Purification and molecular characterization of lamb pregastric lipase. Biochem. Biophys. Acta, 1252:321–329.

de Koning, P. J., 1978. Coagulating enzymes in cheese making. *Dairy Industries International* 43 7–12.

de Llano, D.G., A. Rodriguez and P. Cuesta, 1996. Effect of lactic starter cultures on the organic acid composition of milk and cheese during ripening-analysis by HPLC. *Journal of Applied Bacteriology*, **80**, 570–576.

Emmons, D.B., and M.R. Binns, (1991). Milk clotting enzymes. Proteolysis during cheddar cheese making in relation to estimated losses of basic yield using chymosin derived by fermentation (*A. niger*) and modified enzymes from *M. Miehei*. *Milchwissenschaft*, **6**, 343–346.

Emmons, D.B., D.C. Beckett, and M. Binns, (1990) Milk-Clotting Enzymes. Proteolysis during cheese making in relation to estimated losses of yield. *Journal of Dairy Science*, **73**, 2007–2015.

Ergönül, B., Pelin Günç Ergönül, A. Kemal Seçkin, 2011. Chemical and textural attributes of Hellim cheese, Mljekarstvo 61 (2), 168-174.

Ethiopian Institute of Agricultural Research (EIAR), 2013. http://www.eiar.gov.et/research-center/3-fedral-research-center/26-Holetta-Agricultural-research-center, Accessed on 23 sept. 2013.

FAO-Livestock sector Brief, 2003.

FAO, (2010). [Internet document] URL http://faostat.fao.org/site/610/default.aspx#ancor. Accessed 10/01/2010.

Fasakin, A. and C. Unokiwedi, 1992. Chemical analysis of fermented cheese obtained from Cow milk and melon. *Nigerian. Journal of Microbiology*, 5: 559-566.

Fekadu Beyene, 1994. Present situation and future aspects of milk production, milk handling and processing of dairy products in Southern Ethiopia. Ph.D. Thesis. Department of Food Science, Agricultural University of Norway. Ås, Norway.

Foltmann, B.F., 1992. General and molecular aspects of rennets. Cheese: chemistry, physics and microbiology, Volume 1: General Aspects (ed. P.F. Fox), pp. 37–68, Chapman & Hall, London.

Fox, P.F. and L. Stepaniak, 1993. Enzymes in cheese technology, *International. Dairy Journal*, Vol. 3, P. 509–530.

Fox, P.F, M.G. Timothy, and P.L.H. Mcsweeney, 2000. Fundamentals of cheese science. An aspen publication. Aspen Publishers, Inc Gaithersburg, Maryland.

Fox, P.F. and P.L.H. McSweeney, 1997. Rennets: their role in milk coagulation and cheese ripening. Microbiology and Biochemistry of Cheese and Fermented Milk (ed. B.A. Law), 2nd edn, pp. 1–49, Blackie Academic & Professional, London.

Fox, P.F., 2003. Cheese. In: Encyclopedia of dairy science. Roginski, H., J. W. Fuquay and P. Fox (eds). Elsevier Science: Academic Press, London, U.K. p 252 – 255.

Frazier, W.C. and D.C. Westhoff, 1988. Food microbiology. Tata-McGraw-Hill Publishers Co., New Delhi, pp: 410.

Fuquay, J. W., P. F. Fox and P. L. H. Mcsweeney, 2010. Encyclopedia of dairy science, 2nd Ed. Academic Press Title, 4, Pp. 281–293. Isbn 978-0- 12-374402-9.

Gaborit, P., A. Menard and F. Morgan, 2001. Impact of ripening strains on the typical flavor of goat cheeses. International. Dairy Journal. 11, 315– 325.

Garg, S.K. and B.N. Johri, 1994. Current trends and future research. Food Reviews International, 10, 313–355.

Getachew Feleke and Gashaw Gedda, 2001. The Ethiopian dairy development policy: a draft policy document. Addis Ababa, Ethiopia: Ministry of Agriculture/ AFRDRD/AFRDT Food and Agriculture Organization/SSFF.

Getachew Felleke, 2003. A review of the small-scale dairy sector, Ethiopia: milk and dairy products, post-harvest losses and food safety in sub-saharan Africa and the Near East, FAO prevention of food losses programme.

Getnet Haile, 2009. Impact of global economic crisis on LDC's economic productive capacities and trade prospects: case study the dairy sector in Ethiopia: UNIDO.

Green, M. L. and A. S. Grandison, 1993. Secondary (non-enzymatic) phase of rennet coagulation and post coagulation phenomena. In Cheese: Chemistry, Physics and Microbiology, vol. 1, General aspects, pp. 101±140 (Ed. P. F. Fox). New York: Elsevier Applied Science.

Guinee, T.P. and M.G. Wilkinson, 1992. Rennet coagulation and coagulants in cheese manufacture. *Journal of the Society of Dairy Technology*, 45, 94–104.

Guinee, T.P. and P.F. Fox, 2004. Salt in cheese: Physical, chemical and biological aspects. In: Fox PF, McSweeney PLH, Cogan TM, and Guinee TP (eds.) Cheese, Chemistry, Physics and Microbiology, Vol.1: General Aspects, 3rd edn., pp. 207–259. Amsterdam: Elsevier Academic Press.

Gunasekaran, S., and C. Ay, 1996. Milk coagulation cut-time determination using ultrasonics. *Journal of Food Process Engineering*, 19, 63-73.

Hamed, A.L., N.A. Elsaify, S.L. Faray and F. Orsi, 1992. Effect of pasteurization and storage conditions on the microbiological, chemical and organoleptic of Domati cheese during pickling. Egypt. *Journal of. Dairy Science.*, 20: 177-190.

Harboe, M. and P. Budtz, 1999. The production, action and application of rennet and coagulants. In: Law BA (ed.) Technology of Cheese making, pp. 33–65. Sheffield: Sheffield University Press.

Harboe M, M.L. Broe, and K.B. Qvist, 2010. The production, action and application of rennet and coagulants. In: Law BA and Tamine AY (eds.) Technology of Cheese making, pp. 98–129. Blackwell Publishing Ltd.

Hynes, E., J.C. Ogier and A. Delacroix-Buchet, 2001. Proteolysis during ripening of miniature washed curd cheeses manufactured with different strains of starter bacteria and a *lactobacillus plantarum* adjunct culture. *International. Dairy Journal*, 11 (8): 587- 597.

Hooydonk, A.C. M. and G. Van den berg, 1988. Control and determination of the curd-setting during cheese making. Bull. IDF 225: 2-10.

Horne, D.S., 1998. Casein interactions: casting light on the black boxes the structure in dairy products. *International Dairy Journal* 8:171-177.

Horn, D.s and J.M. Banks, 2004. Rennet induced coagulation of milk. In Fox PF, Mc Sweeney PLH, Cogan TM, and Guinees TP (eds). Cheese; chemistry, physics and microbiology, Vol 1; general aspects, 3rd edition, pp 40-47, London, Elsevier Academic press.

Hori, T., 1985. Objective measurements of the process of curd formation during rennet treatment of milk by the Hot Wire method. *Journal of Food Science*, 50(4), 911-917.

Hough, G., M.L. Puglieso, R. Sanchez and O. Mendes da Silva, 1999. Sensory and microbiological shelf-life of a commercial ricotta cheese. *Journal of Dairy Science.* 82: 454–459.

Jacob, M., D. Jaros and H. Rohm, 2011. Recent advances in milk clotting enzymes, *International Journal of Dairy Technology*, 63, pp. 14–33.

Jayarao, B.M., S.R. Pillay, A.A. Sawant, D.R. Wolfgang and N.V. Hegde, 2004. Guidelines for monitoring bulk tank milk somatic cell and bacterial counts. *Journal of. Dairy Science.* 87 (10): 3561-3573.

Jeremiah, J. Sheehana, Kathleen O'sullivanb and Timothy P. Guineea, 2004. Effect of coagulant type and storage temperature on the functionality of reduced-fat Mozzarella cheese.

Kaminarides, S., E. Rogoti, and H. Mallatou, 2000. Comparison of the characteristics of Halloumi cheese made from ovine milk, caprine milk or mixtures of these milks. *International Journal of Dairy Technology*, 53(3) 100-105.

Kaminarides, S., P. Stamou, T. Massouras, 2007. Changes of organic acids, volatile aroma compounds and sensory characteristics of Halloumi cheese kept in brine. Food Chemistry 100, 219-225.

Kaminarides, S.E. and E.M. Anifantakis, 1985. Changes in the microbiological burden and the quality of Halloumi cheese kept in brine 10% at 4°C and 20°C. In: modern technology and quality control, Proceedings of the First Panhellenic Food Conference, pp. 384–395. Thessaloniki University of Agriculture, Thessaloniki.

Kassahun Melesse, 2008. Characterization of milk products consumption pattern, preference and compositional quality of milk in Ada'a and Lume woreda of East Shoa zone, Central Ethiopia, M.Sc. Thesis.

Kindstedt, P.S., J.J. Yun, D.M. Barbano, and K.L. Larose, 1995. Mozzarella cheese: impact of coagulant concentration on chemical composition, proteolysis, and functional properties. *Journal of Dairy Science*, 78, 2591–2597.

Kirk, S.R. and R, Sawyer, 1991. Pearson's composition and analysis of foods. 9th edition. Longman Scientific and Technical. U.K.

Kubarsepp, I.M., M. Henno, H. Viinalass, and D. Sabre, 2005. Effect of κ-casein and β-lactoglobulin genotypes on the milk rennet coagulation properties. Agr. Res., 1, 55-64.

Lawless, T.H. and H. Heymann, 1999. Sensory evaluation of food; principles and practice. Kluwer Academic/Plenum Publishers. New York. U.S.A. pp 827.

Lawrence, R.C., 1993. Relationship between milk protein genotypes and cheese yield capacity. In: Factors affecting the yield of cheese. Ed. D.B. Emmons. *International. Dairy Federation. Brussels*, 121-127.

Lelievre, J., 1977. Rigidity modulus as a factor influencing the syneresis of renneted milk gels. *Journal of Dairy Research*, 44, 611–614.

Lo´ pez, M.B., S.B. Lomholt and K.B. Qvist, 1998. Rheological properties and cutting time of rennet gels. Effect of pH and enzyme concentration. *International Dairy Journal*, 8, 289–293.

Lucey, J.A., 2002. The relationship between rheological parameters and whey separation in milk protien gels. *Journal of Dairy Science* 85:281-294.

Lucey, J.A., C. Gorry and P.F. Fox, 1994. Methods for improving the rennet coagulation properties of heated milk. Cheese yield and factors affecting its control, Special Issue 9402, pp. 448–456, International Dairy Federation, Brussels.

Luyten, H., 1988. The Rheological and fracture properties of gouda cheese, PhD Thesis, Wageningen Agricultural University, Wageningen, The Netherlands.

Marth, E.H., 1978. Standared methods for examination of dairy products. 14th edition : 358-360.

Martinez-Cuesta, M.A., E. Salas, A. Radomski and M.W. Radomski, 2001. Matrix metalloproteinase-2 In platelet adhesion to fibrinogen: interactions with nitric oxide. Med. Sci. Monitor.;7:646–651.

Milci, S., A.Z. Goncu, H. Alpkent and A. Yaygın, 2005. Chemical, microbiological and sensory characterization of Halloumi cheese produced from ovine, caprine and bovine milk. *International Dairy Journal* 15 (6-9), 625- 630.

Moatsou, G., A. Hatzinaki, G. Psathas, and E. Anifantakis, 2004. Detection of caprine casein in ovine Halloumi cheese. *International Dairy Journal*, 14, 219–226.

Mogessie Ashenafi, 1992. The microbiology of Ethiopian *Ayib*. In application of biotechnology in traditional fermented foods. National Academy of Science Washington D.C USA.

Moschopoulu, E., 2011. Characteristics of rennet and other enzymes from small ruminants used in cheese production, Small Ruminant Research, 10, No. 1, pp. 188–195.

Norman, F., F. Olson., and E. Mark,. Johnson, 1990. Light cheese products: characteristics and economics. Food Technology, 10, 93-96.

O'Connor, C.B., 1993. Traditional cheese making manual. ILCA (International Livestock Centre for Africa), Addis Ababa, Ethiopia.

O'Connor, C.B., 1994. Rural dairy technology. ILRI training manual 1. ILRI (International livestock research institute), Addis Ababa, Ethiopia.

O'Mahoney, F. and J. Peters, 1987. Sub-Saharan Africa options for smallholder milk processing. Food and Agriculture Organization of the United Nations (FAO). World animal review, No.62. pp.16-30.

O'Mahony, F., 1988. Rural dairy technology- experience in Ethiopia. ILCA training manual No. 4. Dairy technology unit. International Livestock Center for Africa (ILCA), Addis Ababa, Ethiopia. 64. pp.

Papademas, P., 2006. Halloumi cheese in brined cheeses. Blackwell publishing ltd., London: 117–136.

Papademas, P. and R.K. Robinson, 1998. Halloumi cheese: the product and its characteristics. *International Journal of Dairy Technology*, 51, 98–103.

Papademas, P. and R.K. Robinson, 2000. Comparison of the chemical, microbiological and sensory characteristics of different types of Halloumi cheese. *International Dairy Journal*, 10, 761–768.

Papademas, P., J. Norman, and R. Robinson, 2000. Properties of full fat, less fat and reduced fat Halloumi cheeses made from bovine milk. Aust. Dairy Foods 22, 30–32.

Papademas, P., 2000. Halloumi cheese: the product and its characteristics. PhD Thesis, School of Food Biosciences, University of Reading.

Payne, F.A., C.L. Hicks and P.S. Shen, 1993. Predicting optimal cutting time of coagulating milk using diffuse relectance. *Journal of Dairy Science*, 76, 48-61.

Pezeshki, A., J. Hesari, A. Ahmadi Zonoz, and B. Ghambarzadeh, 2011. Influence of Withania coagulans protease as a vegetable rennet on proteolysis of Iranian UF white cheese *Journal of. Agricultural. Science. Technology*. Vol. 13: 567-576.

Pometto, A., K. Shetty, G. Paliyath and R. E. Levin, 2006. Food biotechnology, 2nd Ed, Taylor & Francis Group, Llc, 2008 P. Isbn 978-1-4200- 2797-6.

Quade, H.D and H. R¨udiger, 1998. Ausbeutestudie zur K¨aseherstellung. Deutche Milchwirtschaft, 12, 484–487.

Raheem, B., 2006. Developments and microbiological applications in African foods: Emphasis on Nigerian wara cheese. Department of Applied Chemistry and Microbiology Division of Microbiology University Of Helsinki, Finland. ISBN 952-10-3336-3.

Ramet, J.P., 1997. The agents of milk processing. In: The Cheese, Eck, A. and J.C. Gillis (Eds). 3rd Edn. Technical document Publisher, Paris, France, PP: 165-174.

Rampilli, M., L. Raul and H. Marianne, 2005. Natural heterogeneity of chymosin and pepsin in extracts of bovine stomachs. *International Dairy Journal*, March, Vol. 15, No. 11, P. 1130-1137

Raphaelides, S.N., K.D. Antoniou, S. Vasilliadou, C. Georgaki and A. Gravanis, 2006. Ripening effects on the rheological behavior of Halloumi cheese. *Journal of Food Engineering* 76, 321-326.

Repelius, C., 1993. Application of maxiren in cheese making. Milk Proteins '93 (ed. J. Mathieu), pp. 114–121, ENILV, La Roche-sur-Foron, France.

Richardson, G.H., 1985. Standard methods for examination of dairy products. American Public Health Association, Washington D.C.

SAS, 2005. SAS release 9.1. Institute Inc., Cary, NC, USA.

Schmidt, K., and J. Bouma, 1992: Estimating shelf-life of cottage cheese using hazard analysis. *Journal of Dairy Science*, **75**: 2922–2927.

Spangler, P.L., L.A. Jensen, C.H. Amundson, N.F. Olson, and C.G.J. Hill, 1991. Ultrafiltered Gouda cheese: effects of pre acidification, diafiltration, rennet and starter concentration, and time to cut. *Journal of Dairy Science*, **74**, 2809–2819.

Taye Tolemariam, 1998. Qualities of cow milk and the effect of lactoperoxidase system on preservation of milk at Arsi, Ethiopia. M.Sc. Thesis, Alemaya University, Ethiopia.135 pp.

Teply, M., J. Mašek and J. Havlova, 1978. Rennets of animals and microbial of origin, 1 Ed, Snl, Prague, 288 P., 04-812-76.

Tesfaye Kumsa, 1992. Smallholder dairy in Ethiopia. In: Katagile, J. A. and Mubi, S. (eds) Future of livestock industries in East and Southern Africa. Proceedings of a workshop held at Kadoma Ranch Hotel, Zimbabwe, 20-23 July, 1992. ILCA (International Livestock Centre for Africa), Addis Ababa, Ethiopia. pp. 51-58.

Tesfaye Kumsa, 1995. On-farm dairy performance of F1 crossbred cows at highland and mid altitude zones. In: Proceedings of the third National Conference of the Ethiopian Society of Animal Production. 27-29 April 1995. Addis Ababa, Ethiopia.

Toufeili, I., and B. Ozer, 2006. Brined cheeses from the Middle East and Turkey In Brined Cheeses A. Y. Tamime. Blackwell Publishing.

Traordinary, D., 2001. Improving cheese quality: Researching the origin and control of common defects. Available from www.extraordinary dairy.com. Accessed on 20th Oct 2006.

Tsehay Reda, 2002. Small-scale milk marketing and processing in Ethiopia. In: Rangnekar D. and Thorpe W. (eds). Smallholder dairy production and marketing— Opportunities and constraints. Proceedings of a South–South workshop held at NDDB, Anand, India, 13–16 March 2001. NDDB (National Dairy Development Board), Anand, India, and ILRI (International Livestock Research Institute), Nairobi, Kenya.

Tunick, M.H., D.L. Van Hekken, J. Call, F.J. Molina-Corral and A. Gardea, 2007. Queso Chihuahua: Effects of seasonality of cheese milk on rheology. *International Journal of Dairy Technology* 60 13–21.

Turkoglu, H., Z.G. Ceylan and K.S. Dayisoylu, 2003. The microbiological and chemical Quality of orgu cheese produced in Turkey. Pakistan *Journal of Nutrition* 2 (2): 92-94.

Vairo Cavalli, S., S. Claver, N. Priolo, and C. Natalucci, 2005. Extraction and partial characterization ofa coagulant preparation from *Silybum marianum* flowers. Its action on bovine caseinate. *Journal of Dairy Research*, 72: 271-275.

Walstra, P., 1993. The syneresis of curd. In P. F. Fox (Ed.), Cheese: Chemistry, physics and microbiology. Elsevier Appl. Sci. 1:141–191.

Walstra, P., P. Geurts, A. Noomen, A. Jellema, and M.A.J.S. van Boekel, 1999. Dairy technology–Principles of milk properties and processes. Marcel Dekker, New York.

Wigley, R.C., 1996. Cheese and whey. Industrial Enzymology (eds T. Godfrey & S. West), 2nd edn, pp. 133–154, Macmillan Press, London.

Zelalem Yilma and Inger Ledin, 2000. Milk production, processing, marketing and the role of milk and milk products on smallholder farms income in the central highlands of Ethiopia. In: Proceedings of 8th annual conference of the Ethiopian society of Animal production. (ESAP) 24-26 August 2000, Addis Ababa, pp 139-154.

Zoon, P., T. van Vliet, and P. Walstra, 1994. Rheological properties of rennet-induced skim milk gels. *Netherlands Milk and Dairy Journal* 42: 249–269.

APPENDICES

Appendix I. Analysis of Variance Tables

Table 1. ANOVA for clotting activity among treatments in Soxhlet unit

Source	DF	SS	MS	F
Treatment	2	1566.72	783.36	133.75^{**}
Error	9	52.71	5.58	
Total	11	1619.43		
CV (%)	7.32			

Table 2. ANOVA for clotting strength among treatments in Rennin units

Source	DF	SS	MS	F
Treatment	2	0.069	0.034	83.16^{**}
Error	9	0.003	0.0004	
Total	11			
CV (%)	15.65			

Table 3. ANOVA for clotting time in minutes among treatments

Source	DF	SS	MS	F
Treatment	2	442.88	221.44	158.55^{**}
Error	9	12.57	1.39	

Total	11	455.45		
CV (%)	8.98			

Table 4. Repeated measure for stored rennet pH among treatments

Source	DF	SS	MS	F
Treatment	2	1.72	0.86	249.64**
Error	9	0.03	0.003	
Total	11	1.75		
CV (%)	1.59			

Table 5. Repeated measure for stored curd pH among treatments

Source	DF	SS	MS	F
Treatment	2	1.05	0.52	164.84**
Error	9	0.028	0.003	
Total	11	1.081		
CV(%)	0.97			

Table 6. ANOVA for yield of Halloumi cheese among treatments

Source	DF	SS	MS	F
Treatment	2	204.97	102.48	3.17**
Error	9	291.4	32.37	
Total	11	496.37		

CV (%)   4.43

Table 7. ANOVA for pH of Halloumi cheese among treatments

Source	DF	SS	MS	F
Treatment	2	0.47	0.23	47.52**
Error	9	0.044	0.0049	
Total	11	0.517		
CV (%)	1.17			

Table 8. ANOVA for Acidity of Halloumi cheese among treatments

Source	DF	SS	MS	F
Treatment	2	0.0013	0.0006	23.67**
Error	9	0.0002	0.00003	
Total	11	0.0016		
CV (%)	3.33			

Table 9. ANOVA for fat value of Halloumi cheese among treatments

Source	DF	SS	MS	F
Treatment	2	12.50	6.25	0.35ns
Error	9	162.5	18.05	
Total	11	175.0		
CV (%)	8.88			

Table 10. ANOVA for protein value of Halloumi cheese among treatments

Source	DF	SS	MS	F
Treatment	2	1.74	0.87	0.69ns
Error	9	11.41	1.26	
Total	11	13.15		
CV (%)	6.85			

Table 11. ANOVA for total solid value of Halloumi cheese among treatments

Source	DF	SS	MS	F
Treatment	2	47.52	3.76	1.27ns
Error	9	26.56	2.95	
Total	11	74.08		
CV (%)	3.39			

Table 12. ANOVA for moisture value of Halloumi cheese among treatments

Source	DF	SS	MS	F
Treatment	2	47.52	23.76	8.05**
Error	9	26.56	2.95	
Total	11	74.08		
CV (%)	3.48			

Table 13. ANOVA for ash value of Halloumi cheese among treatments

Source	DF	SS	MS	F
Treatment	2	0.029	0.014	0.01ns
Error	9	20.14	2.23	
Total	11	20.17		
CV (%)	9.54			

Table 14. ANOVA for TBC of Halloumi cheese among treatments

Source	DF	SS	MS	F
Treatment	2	0.026	0.0134	43.88**
Error	9	0.0027	0.0003	
Total	11	0.029		
CV (%)	0.30			

Table 15. ANOVA for YMC of Halloumi cheese among treatments

Source	DF	SS	MS	F
Treatment	2	0.07	0.036	4.81**
Error	9	0.06	0.007	
Total	11	0.14		
CV (%)	2.21			

Table 16. ANOVA for CC of Halloumi cheese among treatments

Source	DF	SS	MS	F
Treatment	2	0.020	0.01	1.24ns
Error	9	0.073	0.008	
Total	11	0.09		
CV (%)	3.52			

Table 17. ANOVA for Appearance of Halloumi cheese among treatments

Source	DF	SS	MS	F
Treatment	2	0.12	0.063	0.20ns
Error	9	2.87	0.31	
Total	11	2.99		
CV (%)	13.39			

Table 18. ANOVA for odur of Halloumi cheese among treatments

Source	DF	SS	MS	F
Treatment	2	0.18	0.09	0.91ns
Error	9	0.92	0.10	
Total	11	1.10		
CV (%)	8.06			

Table 19. ANOVA for taste of Halloumi cheese among treatments

Source	DF	SS	MS	F
Treatment	2	0.60	0.30	4.40**
Error	9	0.62	0.068	
Total	11	1.22		
CV (%)	6.67			

Table 20. ANOVA for consistency of Halloumi cheese among treatments

Source	DF	SS	MS	F
Treatment	2	0.08	0.04	0.14ns
Error	9	2.57	0.28	
Total	11	2.65		
CV (%)	13.19			

Appendix II. Halloumi cheese sensory evaluation score table

Date_____ Time_____

Judge name_____

Please taste and evaluate the three samples from left to right and mark 'X' in the box that best describes how much you like the specific samples. Please drink water between samples.

Ser. No.	Sample code	Appearance	Odor	Taste	Consistency
1					
2					
3					

Scoring points

5 is Excellent   **4** is good   **3** fair   **2** poor   **1** is unacceptable quality

COMMENTS

....

....

....

....

....

....